Engineering Your Future

A Brief Introduction to Engineering

FIFTH EDITION

William C. Oakes, PhD
Purdue University

Les L. Leone, PhD
Michigan State University

CONTRIBUTORS

Pramod Abichandani, PhD
Drexel University

John B. Dilworth, PhD
Western Michigan University

Heidi A. Diefes-Dux, PhD
Purdue University

Ralph E. Flori, PhD
University of Missouri-Rolla

Craig J. Gunn, MS
Michigan State University

Marybeth Lima, PhD
Louisiana State University

Merle C. Potter, PhD
Michigan State University

New York Oxford
OXFORD UNIVERSITY PRESS

Oxford University Press is a department of the University of Oxford.
It furthers the University's objective of excellence in research,
scholarship, and education by publishing worldwide.

Oxford   New York
Auckland   Cape Town   Dar es Salaam   Hong Kong   Karachi
Kuala Lumpur   Madrid   Melbourne   Mexico City   Nairobi
New Delhi   Shanghai   Taipei   Toronto

With offices in
Argentina   Austria   Brazil   Chile   Czech Republic   France   Greece
Guatemala   Hungary   Italy   Japan   Poland   Portugal   Singapore
South Korea   Switzerland   Thailand   Turkey   Ukraine   Vietnam

For titles covered by Section 112 of the US Higher Education Opportunity
Act, please visit www.oup.com/us/he for the latest information about pricing
and alternate formats.

Published by Oxford University Press
198 Madison Avenue, New York, NY 10016
www.oup.com

Oxford is a registered trademark of Oxford University Press.

Library of Congress Cataloging-in-Publication Data
Oakes, William C., 1962–
Engineering your future : a brief introduction to engineering / William C. Oakes, PhD, Purdue
University, Les L. Leone, PhD, Michigan State University; contributors, Pramod
Abichandani . . . [et al.].—Fifth edition.
    pages cm
Includes bibliographical references and index.
ISBN 978-0-19-934805-3
1. Engineering—Vocational guidance. 2. Engineering. I. Leone, Les L. II. Gunn, Craig J.
III. Title.
TA157.O215 2014
620.0023—dc23

                                                                    2013025972

*All references to Excel are understood to be a registered trademark of Microsoft Corporation in the
United States and other countries.*

Printing number: 9 8 7 6 5 4 3 2

Printed in the United States of America
on acid-free paper

Contents

Preface

You can't make an educated decision about what career to pursue without adequate information. *Engineering Your Future* endeavors to give you a broad introduction to the study and practice of engineering. In addition to presenting vital information, we've tried to make it interesting and easy to read as well.

You might find Chapter 2, Engineering Majors, to be a tremendous help to you in determining what areas of engineering sound most appealing to you as you begin your education. Our Profiles of Engineers chapter, available on the Companion Website, may also to be of particular interest to you. The chapter includes information from real people—engineers practicing in the field. They discuss their jobs, their lives, and the things they wish they had known going into the profession.

The rest of the book presents such things as a historical perspective of engineering; some thoughts about the future of the profession; some tips on how best to succeed in the classroom; advice on how to gain actual, hands-on experience; exposure to computer-aided design; and a nice introduction to several areas essential to the study and practice of engineering.

We have designed this book for modular use in a freshman engineering course that introduces students to the field of engineering. Such a course differs in content from university to university. Consequently, we have included many topics, too numerous to cover in one course. We anticipate that several of the topics will be selected for a particular course with the remaining topics available to you for outside reading and for future reference.

As you contemplate engineering, you should consider the dramatic impact engineers have had on our world. Note the eloquent words of American Association of Engineering Societies Chair Martha Sloan, a professor of electrical engineering at Michigan Technological University:

> In an age when technology helps turn fantasy and fiction into reality, engineers have played a pivotal role in developing the technologies that maintain our nation's economic, environmental and national security. They revolutionized medicine with pacemakers and MRI scanners. They changed the world with the development of television and the transistor, computers and the Internet. They introduced new concepts in transportation, power, satellite communications, earthquake-resistant buildings, and strain-resistant crops by applying scientific discoveries to human needs.

Engineering is sometimes thought of as applied science, but engineering is far more. The essence of engineering is design and making things happen for the benefit of humanity.

Joseph Bordogna, President of IEEE, adds:

Engineering will be one of the most significant forces in designing continued economic development and success for humankind in a manner that will sustain both the planet and its growing population. Engineers will develop the new processes and products. They will create and manage new systems for civil infrastructure, manufacturing, communications, health care delivery, information management, environmental conservation and monitoring, and everything else that makes modern society function.

We hope that you, too, will find the field of engineering to be attractive, meaningful, and exciting—one that promises to be both challenging and rewarding, and one that matches well with your skills and interests.

For the instructor's convenience, there is a companion website with ancillary materials (PowerPoint lectures and a test bank). This material may be found at
http://www.oup.com/us/companion.websites/9780199797554/

New to the Fifth Brief Edition

- New chapter featuring nine crucial Excel skills all engineering students should master
- Updated chapter on technical communication skills
- Up-to-date statistics for the job market provide critical data to help inform students' career decisions.

Acknowledgments

The authors are especially grateful to the reviewers whose opinion and comments directly influenced the development of this edition:

Spyros Andreou, Savannah State University
Juan M. Caicedo, University of South Carolina
Matthew Cavalli, University of North Dakota
Rafael Fox, Texas A&M University–Corpus Christi
Nikki Larson, Western Washington University
Jennifer Light, Lewis-Clark State College
Edgar Herbert Newman, Coastal Carolina University
G. Albert Popson, Jr., West Virginia Wesleyan College

Nikki Strader, Ohio State University
Yiheng Wang, Danville Community College
Shuming Zheng, Chicago State University

We would also like to thank those reviewers who provided feedback for previous editions:

Keith Gardiner, Lehigh University
Chris Geiger, Florida Gulf Coast University
Yoon Kim, Virginia State University
Keith Level, Las Positas College
S.T. Mau, California State University at Northridge
John Nicklow, Southern Illinois University at Carbondale
Charles E. Pierce, University of South Carolina
Megan Piccus, Springfield Technical Community College
Ken Reid, Ohio Northern University
Gregory Wight, Norwich University
David Willis, University of Massachusetts at Lowell

—*The Authors*

The History of Engineering

1.1 Introduction

Engineering involves a continuum, where every new innovation stands on the shoulders of those who have gone before—perhaps including yours some day. How do you get the necessary insight to participate? Look to the stories of history.

As we begin, you will notice that there were few engineering innovations in the early years. As time passed, innovations occurred more rapidly. Today, engineering discoveries are made almost daily. The speed with which things now change indicates the urgency of understanding the process of innovation. But if we don't maintain a big-picture awareness regarding our projects, we may be surprised by unintended consequences. History provides a great opportunity to observe the context of before, during, and after some of the greatest engineering problems ever faced—and of those we face today.

A main thing to realize is that history is not about memorizing names and dates. History is about people. The insights you'll gain from stories about how engineers developed everything from kitchen appliances to high-tech industrial equipment can be very motivating. And you'll really be able to relate to the history of bridges, for instance, after you've been assigned to build a model bridge in class yourself!

Engineers are professionals. Professionals are leaders. To lead you need to understand the origins of your profession. The stories of history give you the foundation you need, as you learn from the great innovators and see how they handled all aspects of problem-solving. Such broad learning can greatly help your career and aid your development as a leader in your profession.

Many types of professionals are required to master the history of their trade as part of their degrees. The best professionals keep studying throughout their careers. Reading papers, magazines, and journals is simply contemporary "historic" study. This habit builds on their foundational knowledge of history. Such a background gives professionals their best chance to know what it takes to move forward in their field—an amazing challenge, as the stories to come will show!

The study of history, of course, not only helps us create new futures, but it also helps us understand what good qualities from the past are worth emulation. Craftsmanship, integrity, and dedication are clearly evident in our forebears' engineering artifacts. And history is full of interesting educational stories, characters, and ingenious development.

You can learn what it means to do quality work in a quality way, no matter what the level of your contribution to the profession.

Definition of Engineering

Even if you already have a general knowledge of what engineering involves, a look at the definition of the profession may give you more insight. ABET—The Accreditation Board for Engineering and Technology—defines engineering as:

> *The profession in which knowledge of the mathematical and natural sciences, gained by study, experience, and practice, is applied with judgment to develop ways to use, economically, the materials and forces of nature for the benefit of mankind.*

In simple terms, engineering is about using natural materials and forces for the good of mankind—a noble endeavor. This definition places three responsibilities on an engineer: (1) to develop judgment so that you can (2) help mankind in (3) thrifty ways. Looking at case *histories* and *historic* overviews might help us achieve the insight needed to fulfill those responsibilities.

An engineering professor once said that the purpose of an engineer is *"to interpret the development and activity of man."* Technical coursework teaches us skills, but history can teach us how to *interpret* scenarios, and to sort out the pros and cons of various options. History helps us forge bonds of fellowship that connect us to the past and inspire us to be our best for tomorrow. A solid knowledge of history transforms our schooling from training to true education.

1.2 Getting Started

As we proceed, you should note that the engineer has always had a monumental impact on the human race at every stage of societal development. The few items mentioned here are only the tip of the iceberg when it comes to the contributions that engineers have made to the progress of humanity.

Prehistoric Culture

If you look back at the definition of engineering given by ABET, you will notice an important statement: "The profession in which knowledge of the mathematical and natural sciences . . . is applied. . . ." Individuals involved during prehistoric times in activities which we recognize today as engineering—problem-solving, tool-making, etc.—did not have a grasp of mathematical principles nor knowledge of natural science *as we know it today.* They designed and built needed items by trial and error and intuition. They built some spears that worked and some that failed, but in the end they perfected weapons that allowed them to bring down game animals and feed their families. Since written communication and transportation did not exist at that time, little information or innovation was exchanged with people from faraway places. Each group inched ahead on its own.

However, the innovators of yore would have made fine engineers today. Even in light of their limited skill, their carefully cultivated knowledge of their surroundings was more extensive than we typically can comprehend. Their skill in craftsmanship was often marvelous in its effectiveness, integrity, and intricacy. They passed on knowledge of all aspects of life, which they typically treated as a whole entity, with utmost seriousness to the next generation. This information was carefully memorized and kept accurate, evolving, and alive. Early man even tried to pass on vital information by way of coded cave paintings and etchings as an extra safeguard. Breakthroughs in transportation and exploration are being located ever earlier as we continue to make discoveries about various peoples traveling long before we thought they did—influencing others and bringing back knowledge. We can still see something of the prehistoric approach in some of today's cultures, such as Native Americans, aborigines, and others.

However, despite the strong qualities of prehistoric man—the importance placed on respect for life and a sense of the sacred—their physical existence was harshly limited. The work of engineers can be seen in this light as a quest to expand outer capacity without sacrificing inner integrity. The physical limitations of prehistoric man can perhaps be highlighted as follows. (How does engineering impact each of the areas listed?)

Physical limitations of prehistoric cultures:

- They had no written language.
- Their verbal language was very limited.
- They had no means of transportation.
- They had no separate concept of education or specialized methodology to discover new things.
- They lived by gathering food and trying to bring down game with primitive weapons.
- Improvement of the material aspects of life came about very slowly, with early, primitive engineering.

Our Computer Age

You, on the other hand, live in the information age. With such tools as the Internet, the answers to millions of questions are at your fingertips. We have traveled to the moon and our robots have crawled on Mars. Our satellites are exploring the ends of our known universe. Change is no longer questioned; it is expected. We excitedly await the next model in a series, knowing that as the current model is being sold it has already become obsolete. Speed, furious activity, and the compulsion to never sit still are part of our everyday lives. Let's look at the times when constant change was not the norm. As we progress from those primitive times into the 20th century, you will observe that fury of engineering activity.

We will present a panoramic view of engineering by briefly stating some of the more interesting happenings during specific time periods. Notice the kinds of innovations that were introduced. Take a careful look at the relationships that many inventions had with each other. Think about the present, and the connectivity between all areas of engineering and the critical importance of the computer. Innovations do not happen in a vacuum; they are interrelated with the needs and circumstances of the world at the time.

ACTIVITY 1.1   Prepare a report that focuses on engineering in one of the following eras. Analyze the events that you consider to be engineering highlights and explain their importance to the progress of man.

a) Prehistoric man
b) Egypt and Mesopotamia
c) Greece and Rome
d) Europe in the Middle Ages
e) Europe in the Industrial Revolution
f) The 20th century

The assignment above might seem overwhelming—covering an enormous amount of time and information. Actually, the key developments might be simpler to identify than you think. In the pages that follow, however, we will only set the stage for your investigation of your profession. It will be your job to fill in the details for your particular discipline.

The Pace of History

The rate of innovation brings up some interesting points. As we move through the past 6000 years, you should realize that the rate at which we currently introduce innovations is far more rapid than in the past. It used to take years to accomplish tasks that today we perform in a very short time—tasks that we simply take for granted. Think about the last time that your computer was processing slower than you thought it should. Your words may be echoing now: "Come on! Come on! I don't have all day!" In the past, there were often decades without noticeable technological progress. Think of the amount of time that it takes to construct a building today with the equipment that has been developed by engineers. It is not uncommon to see a complete house-frame constructed in a single day. Look in your history books and read about the time it took to create some of the edifices in Europe. You can visit churches today that took as long as 200 years to construct. Would we ever stand for that today?

Now, shift gears and evaluate the *purpose* behind the monumental efforts of the "slow" past. Was their only goal "to get the job done"? Ask them! Look into their stories. There you'll find coherent, colorful explanations for the case of keeping all aspects of culture connected to the main goals of life. The proper connection to God, truth, justice, fate, reality, life, and ancestry was the goal of early science and of many cultures. Not much was allowed to interfere. Even so, the ancients accomplished fantastic feats with only a rudimentary knowledge of the principles you learned as a child. But do we know what else they knew? Perhaps their lack of physical speed was partly voluntary! Perhaps it was surpassed by strengths and insights in other areas. Perhaps today's prowess comes at the loss of other qualities. We need to study history so that we can avoid making Faustian bargains. (Hey! Faust is a character from literary history!) Archimedes, for instance, refused to release information that could be used to make more effective weapons; he knew it would be used for evil and not the pursuit of wisdom; only when his home city of Syracuse was no longer able to hold off Roman attackers did he release his inventions to the military.

The study of history confronts us with dilemmas. Neither side of a true dilemma ever goes away. The story of history is never over. Speed is relative, after all!

Quick Overview

Let's begin with a quick review of the history of engineering—six thousand years in a single paragraph . . .

Our technological roots can be traced back to the seed gatherer/hunter. Prehistoric man survived by collecting seeds and killing what animals he could chase down. He endured a very lean physical existence. As he gradually improved his security through innovation, humanity increased in numbers, and it was important to find ways to feed and control the growing population. To support larger populations reliably, the methods and implements of farming and security were improved. Much later, after many smaller innovations, specialized industrial man stepped onto the scene, ready to bring the world productivity and material wealth. The pursuit of science prospered due to its usefulness and profitability. Much abstract research was done with surplus funding. The microchip was developed. And today, the resulting Technological/Post-Industrial/Computer-Information Man is **you**—ready to use the vast information available in the present day to build the world of the future.

Most innovation would not have come to pass were it not for the work of engineers. The sections that follow will present a brief look at some of the highlights of those 6000 years. Spend some time poring over a few classic history texts to get the inside stories on the innovations that interest you the most. Though you might have interest in one particular field of engineering, you might find stories of innovation and discovery in other disciplines to be equally inspiring.

1.3   **The Beginnings of Engineering**

The Earliest Days

The foundations of engineering were laid with our ancestors' effort to survive and to imp rove their quality of life. From the beginning they looked around their environment and saw areas where life could be made easier and more stable. They found improved ways to hunt and fish. They discovered better methods for providing shelter for their families. Their main physical concern was day-to-day survival. As life became more complicated and small collections of families became larger communities, the need grew to look into new areas of concern: power struggles, acquisition of neighboring tribes' lands, religious observances. All of these involved work with tools. Engineering innovations were needed to further these interests. Of course, in those days projects weren't thought of as separate from the rest of life. In fact, individuals weren't generally thought of as being separate from their community. They didn't look at life from the point of view of specialties and individual interests. Every person was an engineer to an extent.

Modern aborigines still live today much as their ancestors did in prehistoric times. However, frequently even they take advantage of modern engineering in the form of tools, motors and medicine. The Amish can also fall into this category of being a roots-oriented culture. Such cultures tend to use tools only for physical necessities so that the significance of objects doesn't pollute their way of life. This struggle, as we know from contemporary "historic" media reports, has only been partly successful, and has caused conflict and misunderstanding on occasion.

Egypt and Mesopotamia

As cities grew and the need for addressing the demands of the new fledgling societies increased, a significant change took place. People who showed special aptitude in certain areas were identified and assigned to ever more specialized tasks. This labeling and grouping was a scientific breakthrough. It gave toolmakers the time and resources to dedicate themselves to building and innovation. This new social function created the first real engineers, and for the first time innovation flourished rapidly.

Between 4000 and 2000 B.C., Egypt and Mesopotamia were the focal points for engineering activity. Stone tools were developed to help man in his quest for food. Copper and bronze axes were perfected through smelting. These developments were not only aimed at hunting. The development of the plow was allowing man to become a farmer so that he could reside in one place and leave the nomadic life. Mesopotamia also made its mark on engineering by giving birth to the wheel, the sailing boat, and methods of writing. Engineering skills that were applied to the development of everyday items immediately improved life as they knew it. We will never be able to understand completely the vast importance of the Greeks, Romans, and Egyptians in the life of the engineer.

During the construction of the pyramids (c.2700–2500 B.C.) the number of engineers required was immense. They had to make sure that everything fit correctly, that stones were properly transported long distances, and that the tombs would be secure against robbery. Imhotep (chief engineer to King Zoser) was building the Step Pyramid at Sakkara (pictured in Fig. 1.1) in Egypt about 2700 B.C. The more elaborate Great Pyramid of Khufu (pictured in Fig. 1.2) would come about 200 years later. The story of the construction of the pyramids is one that any engineer would appreciate, so consider doing some research on the pyramids on your own. Or perhaps some day you'll have a chance to take a trip to Egypt to see them for yourself. By investigating the construction of the pyramids, you will receive a clear and fascinating education about the need for designing, building, and testing with any engineering project. These early engineers, using simple tools, performed with great acuity, insight, and technical rigor, tasks that even today give us a sense of pride in their achievements.

The Great Pyramid of Khufu (pictured in Fig. 1.2) is the largest masonry structure ever built. Its base measures 756 feet on each side. The 480-foot structure was constructed of over 2.3 million limestone blocks with a total weight of over 58,000,000 tons. Casing blocks of fine limestone were attached to all four sides. These casing stones, some weighing as much as 15 tons, have been removed over the centuries for a wide variety of other uses. It is hard for us to imagine the engineering expertise needed to

Figure 1.1 The Step Pyramid of Sakkara.

Figure 1.2 The Great Pyramid of Khufu.

quarry and move these base and casing stones, and then piece them together so that they would form the pyramid and its covering.

1.4 An Overview of Ancient Engineering

The following sections will give you a feel for what was going on from 2000 B.C. to the present. In this section we will review one specific engineering feat of each of the cultures of ancient Greece, Rome, and China. These overviews are meant to demonstrate the effort that went into engineering activities of the past. They represent only a small portion of the many developments of the time. It is important for you to look closely at what was being accomplished and the impact it had upon the people of the time. We will mention a few of those accomplishments, but it is important that you consider on your own what good came from the activities of those early engineers. How did their innovations set the groundwork for what was to come?

Engineering the Temples of Greece

The Parthenon, shown in Fig. 1.3, was constructed by Iktinos in Athens in 447 B.C. and was completed by 438 B.C. The temple as we know it was to be built on the foundation of a previous temple. The materials that were used came from the salvaged remains of the previous temple. The Parthenon was designed to house a statue of Athena, which was to be carved by Phidias and stand almost forty feet tall. The temple was to make the statue seem proportioned relative to the space within which it was to be housed. Iktinos performed the task that he was assigned, and the temple exists today as a monument to engineering capability.

Figure 1.3 The Parthenon in Athens.

Remember that this was the year 438 B.C. If we asked a contractor today about the skill needed to build this structure with tools that were available 2500 years ago, we'd get quite a reply.

The Roman Roads and Aqueducts

Construction of the first great Roman Road, the Appian Way, began around 312 B.C. It connected Rome and Capua, a distance of 142 miles. The Appian Way eventually stretched to Brundisium at the very southernmost point in Italy, and covered 360 miles. With this start, the Roman engineers continued building roads until almost A.D. 200. Twenty-nine major roads eventually connected Rome to the rest of the empire. By A.D. 200 construction ceased, and repair and maintenance were the only work done on these roads. Aqueducts were part of this construction. One such aqueduct is shown in Fig. 1.4.

The Great Wall of China

In 220 B.C. of the Ch'in Dynasty, Meng T'ien, a military general, led his troops along the borders of China. His primary role was that of a commander of troops charged with the task of repelling the nomadic hordes of Mongolians who occasionally surged across the Chinese border. The Ch'in emperor, Shih Huang Ti, commissioned him to begin the building of what would become known as the Great Wall of China; see Fig. 1.5.

Figure 1.4 Roman Aqueduct.

Figure 1.5 The Great Wall of China.

1.5 **Traveling Through the Ages**

At the same time that the previous monuments were being built, there were a number of other engineering feats under way. Let's look at a few of those other undertakings.

As you read through the following, we encourage you to investigate on your own any of the engineering accomplishments which grab your attention from this list. You should get an overall sense of both the pace and the focus of development over the centuries from these lists.

Note that as we enter the modern age, the scope of invention appears to narrow, with much of the activity relating to computers. What impact does this have on the notion of "ever-faster historic development"?

What is the role of the inventor in history? Sometimes the name is important, other times not. At times during certain eras, innovations were being made simultaneously by a number of people. So no individual really stands out. Perhaps the developments were more collective during such times. Perhaps the players involved were often racing each other to the patent office. At other times, with certain inventions, a single person made a significant breakthrough on his own.

When is it the person and when is it the times? Here are a few clues. When it's the person, his peers might think he's a nut, his work might even be outlawed or ignored. When it's the times, there are frequently many innovators doing similar work in close proximity or even in cooperation. Sometimes even when a name stands out, you still get the impression that his effort was more communal than singular. These are concepts to consider as you watch time and innovations flow past in these lists and in your studies.

1200 B.C.–A.D. 1

- The quality of wrought iron is improved.
- Swords are mass produced.
- Siege towers are perfected.
- The Greeks develop manufacturing.
- Archimedes introduces mathematics in Greece.
- Concrete is used for the arched bridges, roads, and aqueducts in Rome.

ACTIVITIES 1.2

a) Investigate the nature of manufacturing in these early times.

b) Investigate warfare as it was first waged. What was the engineer's role in the design of war-related equipment? What did these engineers build during peacetime?

c) Concrete was being used in Rome before A.D. 1. Trace the history of concrete from its early use to its prominence in building today.

d) Metals have always been an important part of the history of engineering. Investigate the progression in the use of metals from copper and bronze to iron and steel. What effect did the use of these metals have on the societies in which they were used?

A.D. 1–1000

- The Chinese further develop the study of mathematics.
- Gunpowder is perfected.
- Cotton and silk are manufactured.

1000–1400

- There is growth in the silk and glass industries.
- Leonardo Fibonacci (1170–1240), medieval mathematician, writes the first Western text on algebra.

ACTIVITY 1.3   What influence does the introduction of the first Western algebra text have on the society of the time? Why would engineers be interested in this text?

1400–1700

- Georgius Agricola's *De re metallica*, a treatise on mining and metallurgy, is published posthumously.
- Federigo Giambelli constructs the first time bomb for use against Spanish forces besieging Antwerp, Belgium.
- The first water closet (toilet) is invented in England.
- Galileo begins constructing a series of telescopes, with which he observes the rotation of the sun and other phenomena supporting the Copernican heliocentric theory.
- Using dikes and windmills, Jan Adriaasz Leeghwater completes drainage of the Beemstermeer, the largest project of its kind in Holland (17,000 acres).
- Otto von Guericke, mayor of Magdeburg, first demonstrates the existence of a vacuum.
- Christian Huygens begins work on the design of a pendulum-driven clock.
- Robert Hooke develops the balance spring to power watches.
- Charles II charters the Royal Society, England's first organization devoted to experimental science.
- Isaac Newton constructs the first reflecting telescope.

- Work is completed on the Languedoc Canal, the largest engineering project of its kind in Europe.
- Thomas Savery patents his "miner's friend," the first practical steam pump.
- The agriculture, mining, textile, and glassmaking industries are expanded.
- The concept of the scientific method of invention and inquiry is originated.
- The humanities and science are first thought to be two distinctly separate entities.
- Robert Boyle finds that gas pressure varies inversely with volume (Boyle's Law).
- Leibniz makes a calculating machine to multiply and divide.

1700–1800

- The Leyden jar stores a large charge of electricity.
- The Industrial Revolution begins.
- James Watt makes the first rotary engine.
- The instrument-maker Benjamin Huntsman develops the crucible process for manufacturing steel, improving quality and sharply reducing cost.
- Louis XV of France establishes the Ecole des Ponts et Chausses, the world's first civil engineering school.
- John Smeaton completes construction of the Eddystone lighthouse.
- James Brindley completes construction of the Bridgewater Canal, beginning a canal boom in Britain.
- James Watt patents his first steam engine.
- The spinning jenny and water frame, the first successful spinning machines, are patented by James Hargreaves and Richard Arkwright, respectively.
- Jesse Ramsden invents the first screw-cutting lathe, permitting the mass production of standardized screws.
- The Society of Engineers, Britain's first professional engineering association, is formed in London.
- David Bushnell designs the first human-carrying submarine.
- John Wilkinson installs a steam engine to power machinery at his foundry in Shropshire, the first factory use of the steam engine.

- Abraham Darby III constructs the world's first cast iron bridge over the Severn River near Coalbrookdale.
- Claude Jouffroy d'Abbans powers a steamboat upstream for the first time.
- Joseph-Michel and Jacques-Etienne Montgolfier construct the first passenger-carrying hot air balloon.
- Henry Cort patents the puddling furnace for the production of wrought iron.
- Joseph Bramah designs his patent lock, which remains unpicked for 67 years.

- British civil engineer John Rennie completes the first building made entirely of cast iron.

**ACTIVITIES
1.4**

a) The Industrial Revolution changed the whole landscape of the world. How did the engineer fit into this revolution? What were some of the major contributions?

b) Research and compare the first rotary engine to the rotary engines of today.

1800–1825

- Automation is first used in France.
- The first railroad locomotive is unveiled.
- Jean Fourier, French mathematician, states that a complex wave is the sum of several simple waves.
- Robert Fulton begins the first regular steamboat service with the *Clermont* on the Hudson River in the U.S.
- Chemical symbols as they are used today are developed.
- The safety lamp for protecting miners from explosions is first used.
- The single wire telegraph line is developed.
- Photography is born.
- Electromagnetism is studied.
- The thermocouple is invented.
- Aluminum is prepared.
- Andre Ampere shows the effect of electric current in motors.
- Sadi Carnot finds that only a fraction of the heat produced by burning fuel in an engine is converted into motion. This forms the basis of modern thermodynamics.

1825–1875

- Rubber is vulcanized by Charles Goodyear in the United States.
- The first iron-hulled steamer powered by a screw propeller crosses the Atlantic.
- The rotary printing press comes into service.
- Reinforced concrete is used.
- Isaac Singer invents the sewing machine.
- George Boole develops symbolic logic.
- The first synthetic plastic material— celluloid—is created by Alexander Parkes.

- Henry Bessemer originates the process to mass-produce steel cheaply.
- The first oil well is drilled near Titusville, Pennsylvania.
- The typewriter is perfected.
- The Challenge Expedition (1871–1876) forms the basis for future oceanographic study.

1875–1900

- The telephone is patented in the United States by Alexander Graham Bell.
- The phonograph is invented by Thomas Edison.
- The incandescent lightbulb also is invented by Edison.
- The steam turbine appears.
- The gasoline engine is invented by Gottlieb Daimler.
- The automobile is introduced by Karl Benz (Fig. 1.6).

1900–1925

- The Wright brothers complete the first sustained flight.
- Detroit becomes the center of the auto industry.
- Stainless steel is introduced in Germany.
- Tractors with diesel engines are produced by Ford Motor Company.
- The first commercial airplane service between London and Paris commences.
- Diesel locomotives appear.

1925–1950

- Modern sound recordings are introduced.
- John Logie Baird invents a high-speed mechanical scanning system, which leads to the development of television.

Figure 1.6 Karl Benz's three-wheeler automobile.
Source: © Library of Congress - digital ve/Science Faction/Corbis

- The Volkswagen Beetle goes into production.
- The first nuclear bombs are used.
- The transistor is invented.

1950–1975

- Computers first enter the commercial market.
- Computers are in common use by 1960.
- The first artificial satellite—Sputnik 1, USSR—goes into space.
- Explorer I, the first U.S. satellite, follows.
- The laser is introduced.
- Manned space flight begins.
- The first communication satellite—Telstar—goes into space.
- Integrated circuits are introduced.
- The first manned moon landing occurs.

1975–1990

- Supersonic transport from U.S. to Europe begins.
- Cosmonauts orbit the earth for a record 180 days.
- The Columbia space shuttle is reused for space travel.
- The first artificial human heart is implanted.

1990–Today

- Robots walk on Mars.
- Computer processor speed is dramatically improved.
- The Channel Tunnel (the "Chunnel") between England and France is completed.
- World's new tallest building opens in Kuala Lumpur, Malaysia (1,483 feet).
- Global Positioning Satellite (GPS) technology is declassified, resulting in hundreds of safety, weather and consumer applications.

You are the potential innovator of tomorrow. Research the progress made over the past decade in engineering on your own. Consider possible engineering innovations to which you might one day contribute.

EXERCISES AND ACTIVITIES

1.1　The history of engineering is long and varied. It contains many interesting inventions and refinements. Select one of these inventions and discuss the details of its creation. For example, you might explain how the first printing presses came into being and what previous inventions were used to create the new device.

1.2　Build a simple model of one of the inventions mentioned in this chapter—a bridge, aqueduct, or submarine, for instance. Explain the difficulties of building these devices during the time they were invented.

1.3　Explain how easy it would be to create some inventions of the past using our present-day knowledge and capability.

1.4　Engineering history is filled with great individuals who have advanced the study and practice of engineering. Investigate an area of engineering that is interesting to you and write a detailed report on an individual who made significant contributions in that area.

1.5　Explain what it would have been like to have been an engineer during any particular historical era.

1.6　Compare the lives of any two engineers from the past. Are there similarities in their experiences, projects, and education?

1.7　What kind of education were engineers of old able to obtain?

1.8　What period of engineering history interests you most? Why? Explain why this period is so important in the history of engineering.

1.9　If you had to explain to a 7-year-old child why engineering is important to society, what information from the history of engineering would you relate? Why?

CHAPTER 2

Engineering Majors

2.1 Introduction

Engineers produce things that impact us every day. They invent, design, develop, man-ufacture, test, sell, and service products and services that improve the lives of people. The Accreditation Board for Engineering and Technology (ABET), which is the national board that establishes accreditation standards for all engineering programs, defines en-gineering as follows [Landis]:

> *Engineering is the profession in which a knowledge of the mathematical and natu-ral sciences, gained by study, experience, and practice, is applied with judgment to develop ways to utilize, economically, the materials and forces of nature for the benefit of mankind.*

Frequently, students early in their educational careers find it difficult to understand exactly what engineers do, and often more to the point, where they fit best in the vast array of career opportunities available to engineers.

Common reasons for a student to be interested in engineering include:

1. Proficiency in math and science
2. Suggested by a high school counselor
3. Has a relative who is an engineer
4. Heard it's a field with tremendous job opportunity
5. Read that it has high starting salaries

While these can be valid reasons, they don't imply a firm understanding of engineer-ing. What is really important is that a student embarking upon a degree program, and ultimately a career, understands what that career entails and the options it presents. We all have our own strengths and talents. Finding places to use those strengths and talents is the key to a rewarding career.

The purpose of this chapter is to provide information about some of the fields of engineering in order to help you decide if this is an area that you might enjoy. We'll explore the role of engineers, engineering job functions, and the various engineering disciplines.

The Engineer and the Scientist

To better understand what engineers do, let's contrast the roles of engineers with those of the closely related field of the scientist. Many students approach both fields for similar reasons: they were good at math and science in high school. While this is a prerequisite for both fields, it is not a sufficient discriminator to determine which is the right career for a given individual.

The main difference between the engineer and the scientist is in the object of each one's work. The scientist searches for answers to technological questions to obtain a knowledge of why a phenomenon occurs. The engineer also searches for answers to technological questions, but always with an application in mind.

Theodore Von Karman, one of the pioneers of America's aerospace industry, said, "Scientists explore what is; engineers create what has not been." [Paul Wright, *Introduction to Engineering*].

In general, science is about discovering things or acquiring new knowledge. Scientists are always asking, "Why?" They are interested in advancing the knowledge base that we have in a specific area. The answers they seek may be of an abstract nature, such as understanding the beginning of the universe, or more practical, such as the reaction of a virus to a new drug.

The engineer also asks, "Why?" but it is because of a problem which is preventing a product or service from being produced. The engineer is always thinking about the application when asking why. The engineer becomes concerned with issues such as the demand for a product, the cost of producing the product, the impact on society, and the environment of the product.

Scientists and engineers work in many of the same fields and industries but have different roles. Here are some examples:

- Scientists study the planets in our solar system to understand them; engineers study the planets so they can design a spacecraft to operate in the environment of that planet.
- Scientists study atomic structure to understand the nature of matter; engineers study the atomic structure in order to build smaller and faster microprocessors.
- Scientists study the human neurological system to understand the progression of neurological diseases; engineers study the human neurological system to design artificial limbs.
- Scientists create new chemical compounds in a laboratory; engineers create processes to mass-produce new chemical compounds for consumers.
- Scientists study the movement of tectonic plates to understand and predict earthquakes; engineers study the movement of tectonic plates to design safer buildings.

The Engineer and the Engineering Technologist

Another profession closely related to engineering is engineering technology. Engineering technology and engineering have similarities, yet there are differences; they have different career opportunities. ABET, which accredits engineering technology programs as well as engineering programs, defines engineering technology as follows:

Engineering technology is that part of the technological field which requires the application of scientific and engineering knowledge and methods combined with technical skills in support of engineering activities; it lies in the occupational spectrum between the craftsman and engineering at the end of the spectrum closest to the engineer.

Technologists work with existing technology to produce goods for society. Technology students spend time in their curricula working with actual machines and equipment that are used in the jobs they will accept after graduation. By doing this, technologists are equipped to be productive in their occupation from the first day of work.

Both engineers and technologists apply technology for the betterment of society. The main difference between the two fields is that the engineer is able to create new technology through research, design, and development. Rather than being trained to use specific machines or processes, engineering students study additional mathematics and engineering science subjects. This equips engineers to use these tools to advance the state of the art in their field and move technology forward.

There are areas where engineers and engineering technologists perform very similar jobs. For example, in manufacturing settings, engineers and technologists are employed as supervisors of assembly line workers. Also, in technical service fields both are hired to work as technical support personnel supporting equipment purchased by customers. However, most opportunities are different for engineering and engineering technology graduates.

- The technologist identifies the computer networking equipment necessary for a business to meet its needs and oversees the installation of that equipment; the engineer designs new computer boards to transmit data faster.
- The technologist develops a procedure to manufacture a shaft for an aircraft engine using a newly developed welding technique; the engineer develops the new welding machine.
- The technologist analyzes a production line and identifies new robotic equipment to improve production; the engineer develops a computer simulation of the process to analyze the impact of the proposed equipment.
- The technologist identifies the equipment necessary to assemble a new CD player; the engineer designs the new CD player.
- The technologist identifies the proper building materials and oversees the construction of a new building; the engineer determines the proper support structures, taking into account the local soil, proposed usage, earthquake risks and other design requirements.

What Do Engineers Do?

Engineering is an exciting field with a vast range of career opportunities. In trying to illustrate the wide range of possibilities, Professors Jane Daniels and Richard Grace from Purdue University constructed the cubic model of Figure 2.1. One edge of the cube represents the engineering disciplines that most students identify as the potential majors.

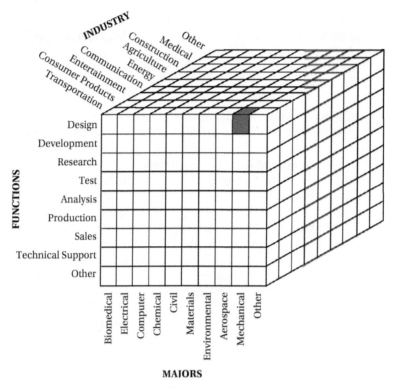

Figure 2.1 Engineering positions. [Grace and Daniels]

A second edge of the cube represents the different job functions an engineer can have within a specific engineering discipline. The third edge of the cube represents industrial sectors where engineers work. A specific engineering position, such as a mechanical engineering design position in the transportation sector, is the intersection of these three axes. As one can see from the cube, there is a vast number of possible engineering positions.

In the following sections of this chapter, the engineering functions and majors are described. The remaining axis, the industrial sectors, are dependent on the companies and governmental agencies that employ engineers.

To obtain more information about the various industrial sectors:

- Explore your school's placement center
- Visit job fairs
- Attend seminars on campus sponsored by various companies
- Search the Internet (visit websites describing career opportunities)
- Talk to faculty familiar with a certain industry
- "Shadow" a practicing engineer
- Work as an intern or co-op engineer
- Take an engineering elective course

Most engineering curricula include common courses within an engineering major for the first two years of study. It is not until the junior or senior year that students take technical electives that can be industry specific. Students are encouraged to explore various career opportunities so they can make better decisions when required to select their junior and senior level electives.

EXAMPLE 2.1

Let's consider a mechanical engineer (an "ME") who performs a design function to illustrate how a specific job in one industrial sector can vary from that in another sector.

- **Aerospace**—*Design of an aircraft engine fan blade*: Detailed computer analyses and on-engine testing are required for certification by the FAA. Reliability, efficiency, cost, and weight are all design constraints. The design engineer must push current design barriers to optimize the design constraints, potentially making tradeoffs between efficiency, cost, and weight.
- **Biomedical**—*Design of an artificial leg and foot prosthesis giving additional mobility and control to the patient*: Computer modeling is used to model the structure of the prosthesis and the natural movement of a leg and foot. Collaboration with medical personnel to understand the needs of the patient is a critical part of the design process. Reliability, durability, and functionality are the key design constraints.
- **Power**—*Design of a heat recovery system in a power plant, increasing the plant productivity*: Computer analyses are performed as part of the design process. Cost, efficiency, and reliability are the main design constraints. Key mechanical components can be designed with large factors of safety since weight is not a design concern.
- **Consumer products**—*Design of a pump for toothpaste for easier dispensing*: Much of the development work might be done with prototypes. Cost is a main design consideration. Consumer appeal is another consideration and necessitates extensive consumer testing as part of the development process.
- **Computer**—*Design of a new ink-jet nozzle with resolution approaching laser printer quality*: Computer analyses are performed to ensure that the ink application is properly modeled. Functionality, reliability, and cost are key design concerns.

2.2 Engineering Functions

Within engineering there are basic classifications of jobs that are common across the various engineering disciplines. What follows are brief descriptions of these different engineering job functions. A few examples are provided for each function. It is important to realize that all the fields of engineering have roles in each of the main functions described here.

Research

The role of the engineering researcher is the closest to that of a scientist of all the engineering functions. Research engineers explore fundamental principles of chemistry, physics, biology, and mathematics in order to overcome barriers preventing advancement in their field. Engineering researchers differ from scientists in that they are interested in the application of a breakthrough, whereas scientists are concerned with the knowledge that accompanies a breakthrough.

Research engineers conduct investigations to extend knowledge using various means. One of the means is conducting experiments. Research engineers may be involved in the design and implementation of experiments and the interpreting of the results. Typically, the research engineer does not perform the actual experiment. Technicians are usually called upon for the actual testing. Large-scale experiments may involve the coordination of additional supporting personnel including other engineers, scientists, technologists, technicians, and craftspeople.

Research is also conducted using the computer. Computational techniques are developed to calculate solutions to complex problems without having to conduct costly and time-consuming experiments. Computational research requires the creation of mathematical models to simulate the natural occurring phenomena under study. Research engineers also might develop the computational techniques to perform the complex calculations in a timely and cost-effective fashion.

Most research engineers work for some type of research center. A research center might be a university, a government laboratory such as NASA, or an industrial research center. In most research positions an advanced degree is required, and often a Ph.D. is needed. If research appeals to you, a great way to explore it is by doing an undergraduate research project with an engineering professor. This will allow you to observe the operation of a laboratory first-hand and find out how well you enjoy being part of a research team.

Development

Development engineers bridge the gap between laboratory research and full-scale production. The development function is often coupled with research in so-called R&D (research and development) divisions. Development engineers take the knowledge acquired by the researchers and apply it to a specific product or application. The researcher may prove something is possible in a laboratory setting; the development engineer shows that it will work on a large, production-size scale and under actual conditions encountered in the field. This is done in pilot manufacturing plants or by using prototypes.

Development engineers are continuously looking for ways to incorporate the findings of researchers into prototypes to test their feasibility for use in tomorrow's products. Often, an idea proven in a laboratory needs to be significantly altered before it can be introduced on a mass production scale. It is the role of development engineers to identify these areas and work with the design engineers to correct them before full-scale production begins.

An example of a development process is the building of concept cars within the automotive industry. These are unique cars that incorporate advanced design concepts and technology. The cars are then used as a test case to see if the design ideas and technology actually perform as predicted. The concept cars are put through exhaustive tests to determine how well the new ideas enhance a vehicle's performance. Each year, new technology is introduced into production automobiles that was first proven in development groups using concept vehicles.

Testing

Test engineers are responsible for designing and implementing tests to verify the integrity, reliability, and quality of products before they are introduced to the public. The test engineer devises ways to simulate the conditions a product will be subjected to during its life. Test engineers work closely with development engineers in evaluating prototypes and pilot facilities. Data from these initial development tests are used to decide whether full production versions will be made or if significant changes are needed before a full-scale release. Test engineers work with design engineers to identify the changes in the product to ensure its integrity.

A challenge that engineers face is simulating the conditions a product will face during its life span, and doing so in a timely, cost-effective manner. Often the conditions the product will face are difficult to simulate in a laboratory. A constant problem for the test engineer is simulating the aging of a product. An example of such a testing challenge is the testing of a pacemaker for regulating a patient's heart which is designed to last several decades. An effective test of this type cannot take 20 years or the product will be obsolete before it is introduced. The test engineer must also simulate conditions within the human body without exposing people to unnecessary risks.

Other challenges facing test engineers involve acquiring accurate and reliable data. The test engineer must produce data that show the product is functioning properly or identify areas of concern. Test engineers develop data acquisition and instrumentation methods to achieve this. Techniques such as radio telemetry may be used to transmit data from the inside of a device being tested. The measurement techniques must not interfere with the operation of the device, presenting a tremendous challenge for small, compact products. Test engineers must cooperate with design engineers to determine how the device being tested can be fitted with instrumentation yet still meet its design intent.

Test engineers must have a wide range of technical and problem-solving skills. They must also be able to work in teams involving a wide range of people. They work with design and development engineers, technicians, and craftspeople, as well as management.

EXAMPLE 2.2 Test engineers must understand the important parameters of their tests. The development of a certain European high-speed train provides an example of the potential consequences which may result when test engineers fail to understand these parameters. A test was needed to show that the windshield on the locomotive could withstand the high-velocity impacts of birds or other objects it might encounter. This is also a common design constraint encountered in airplane design. The train test engineers borrowed a "chicken gun" from an aerospace firm for this test. A chicken gun is a mechanism used to propel birds at a target, simulating in-flight impact. With modern laws governing cruelty to animals, the birds are humanely killed and frozen until the test.

On the day of the test, the test engineers aimed the gun at the locomotive windshield, inserted the bird and fired. The bird not only shattered the windshield but put a hole through the engineer's seat.

The design engineers could not understand what went wrong. They double-checked their calculations and determined that the windshield should have held. The problem became clear after the test engineers reviewed their procedure with the engineers from the aerospace firm from whom they had borrowed the equipment. The aerospace test engineers asked how long they had let the bird thaw.

The response was, "Thaw the bird?"

There is a significant difference in impact force between a frozen eight-pound bird and a thawed bird. The test was successfully completed later with a properly thawed bird.

Design

The design function is what many people think of when they think of engineering, and this is where the largest number of engineers are employed. The design engineer is responsible for providing the detailed specifications of the products society uses.

Rather than being responsible for an entire product, most design engineers are responsible for a component or part of the product. The individual parts are then assembled into a product such as a computer, automobile, or airplane. Design engineers produce detailed dimensions and specifications of the part to ensure that the component fits properly with adjoining pieces. They use modern computer design tools and are often supported by technicians trained in computer drafting software.

The form of the part is also a consideration for the design engineer. Design engineers use their knowledge of scientific and mathematical laws, coupled with experience, to generate a shape to meet the specifications of the part. Often, there is a wide range of possibilities and considerations. In some fields, these considerations are ones that can be calculated. In others, such as in consumer products, the reaction of a potential customer to a shape may be as important as how well the product works.

The design engineer also must verify that the part meets the reliability and safety standards established for the product. The design engineer verifies the integrity of the product. This often requires coordination with analysis engineers to simulate complex products and field conditions, and with test engineers to gather data on the integrity of the product. The design engineer is responsible for making corrections to the design based on the results of the tests performed.

In today's world of ever increasing competition, the design engineer must also involve manufacturing engineers in the design process. Cost is a critical factor in the design process and may be the difference between a successful product and one that fails.

Communication with manufacturing engineers is therefore critical. Often, simple design changes can radically change a part's cost and affect the ease with which the part is made.

Design engineers also work with existing products. Their role includes redesigning parts to reduce manufacturing costs and time. They also work on redesigning products that have not lived up to expected lives or have suffered failure in the field. They also modify products for new uses. This usually requires additional analysis, minor redesigns, and significant communication between departments.

Analysis

Analysis is an engineering function performed in conjunction with design, development and research. Analysis engineers use mathematical models and computational tools to provide the necessary information to design, development, or research engineers to help them perform their function.

Analysis engineers typically are specialists in a technology area important to the products or services being produced. Technical areas might include heat transfer, fluid flow, vibrations, dynamics, system modeling, and acoustics, among others. They work with computer models of products to make these assessments. Analysis engineers often possess an advanced degree and are experienced in their area of expertise.

In order to produce the information required of them, they must validate their computer programs or mathematical models. To do so may require comparing test data to their predictions. This requires coordination with test engineers in order to design an appropriate test and to record the relevant data.

An example of the role of analysis is the prediction of temperatures in an aircraft engine. Material selection, component life estimates, and design decisions are based in large part on the temperature the parts attain and the duration of those temperatures. Heat transfer analyses are used to determine these temperatures. Engine test results are used to validate the temperature predictions. The permissible time between engine overhauls can depend on these temperatures. The design engineers then use these results to ensure reliable aircraft propulsion systems.

Systems

Systems engineers work with the overall design, development, manufacture, and operation of a complete system or product. Design engineers are involved in the design of individual components, but systems engineers are responsible for the integration of the components and systems into a functioning product.

Systems engineers are responsible for ensuring that the components interface properly and work as a complete unit. Systems engineers are also responsible for identifying the overall design requirements. This may involve working with customers or marketing personnel to accurately determine market needs. From a technical standpoint, systems engineers are responsible for meeting the overall design requirements.

Systems engineering is a field that most engineers enter only after becoming proficient in an area important to the systems, such as component design or development. Some graduate work often is required prior to taking on these assignments. However, there are some schools where an undergraduate degree in systems engineering is offered.

Manufacturing and Construction

Manufacturing engineers turn the specifications of the design engineer into a tangible reality. They develop processes to make the products we use every day. They work with diverse teams of individuals, from technicians on the assembly lines to management, in order to maintain the integrity and efficiency of the manufacturing process.

It is the responsibility of manufacturing engineers to develop the processes for taking raw materials and changing them into the finished pieces that the design engineers detailed. They utilize state-of-the-art machines and processes to accomplish this. As technology advances, new processes often must be developed for manufacturing the products.

The repeatability or quality of manufacturing processes is an area of increasing concern to modern manufacturing engineers. These engineers use statistical methods to determine the precision of a process. This is important, since a lack of precision in the manufacturing process may result in inferior parts that cannot be used or that may not meet the customer's needs. Manufacturing engineers are very concerned about the quality of the products they produce. High quality manufacturing means lower costs since more parts are usable, resulting in less waste. Ensuring quality means having the right processes in place, understanding the processes and working with the people involved to make sure the processes are being maintained at peak efficiency.

Manufacturing engineers also keep track of the equipment in a plant. They schedule and track required maintenance to keep the production line moving. They also must track the inventories of raw materials, partially finished parts, and completely finished parts. Excessive inventories tie up substantial amounts of cash that could be used in other parts of the company. The manufacturing engineer is also responsible for maintaining a safe and reliable workplace, including the safety of the workers at the facility and the environmental impact of the processes.

Manufacturing engineers must be able to work with diverse teams of people including design engineers, tradesmen, and management. Current "just in time" manufacturing practices reduce needed inventories in factories but require manufacturing engineers at one facility to coordinate their operation with manufacturing engineers at other facilities. They also must coordinate the work of the line workers who operate the manufacturing equipment. It is imperative that manufacturing engineers maintain a constructive relationship with their company's trade workers.

Manufacturing engineers play a critical role in modern design practices. Since manufacturing costs are such an important component in the success of a product, the design process must take into account manufacturing concerns. Manufacturing engineers identify high cost or high risk operations in the design phase of a product. When problems are identified, they work with the design engineers to generate alternatives.

In the production of large items such as buildings, dams, and roads, the production engineer is called a construction engineer rather than a manufacturing engineer. However, the role of the construction engineer is very similar to that of the manufacturing engineer. The main difference is that the construction engineer's production facility is typically outdoors while the manufacturing engineer's is inside a factory. The functions of construction engineers are the same as mentioned above when "assembly line" and "factory" are replaced with terms like "job site," reflecting the construction of a building, dam, or other large-scale project.

Operations and Maintenance

After a new production facility is brought on-line, it must be maintained. The operations engineer oversees the ongoing performance of the facility. Operations engineers must have a wide range of expertise dealing with the mechanical and electrical issues involved with maintaining a production line. They must be able to interact with manufacturing engineers, line workers and technicians who service the equipment. They must coordinate the service schedule of the technicians to ensure efficient service of the machinery, minimizing its downtime impact on production.

Maintenance and operations engineers also work in non-manufacturing roles. Airlines have staffs of maintenance engineers who schedule and oversee safety inspections and repairs. These engineers must have expertise in sophisticated inspection techniques to identify all possible problems.

Large medical facilities and other service sector businesses require operations or maintenance engineers to oversee the operation of their equipment. It is obviously critical that emergency medical equipment be maintained in peak working order.

Technical Support

A technical support engineer serves as the link between customer and product and assists with installation and setup. For large industrial purchases, technical support may be included in the purchase price. The engineer may visit the installation site and oversee a successful start-up. For example, a new power station for irrigation might require that the technical support engineer supervises the installation and helps the customer solve problems to get the product operational. To be effective, the engineer must have good interpersonal and problem-solving skills as well as solid technical training.

The technical support engineer may also trouble-shoot problems with a product. Serving on a computer company's help line is one example. Diagnosing design flaws found in the field once the product is in use is another example.

Technical support engineers do not have to have in-depth knowledge of each aspect of the product. However, they must know how to tap into such knowledge at their company.

Modern technical support is being used as an added service. Technical support engineers work with customers to operate and manage their own company's equipment as well as others. For example, a medical equipment manufacturer might sell its services to a hospital to manage and operate its highly sophisticated equipment. The manufacturer's engineers would not only maintain the equipment, but also help the hospital use its facilities in the most efficient way.

Customer Support

Customer support functions are similar to those of technical support as a link between the manufacturer and the customer. However, customer support personnel also are involved in the business aspect of the customer relationship. Engineers are often used for this function because of their technical knowledge and problem solving ability. Typically, these positions require experience with the products and customers, and also require some business training.

The customer support person works with technical support engineers to ensure proper customer satisfaction. Customer support is also concerned with actual or perceived value for customers. Are they getting what they paid for? Is it cost-effective to continue current practices? Customer support persons are involved in warranty issues, contractual agreements, and the value of trade-in or credits for existing equipment. They work very closely with the technical support engineers and also with management personnel.

Sales

Engineers are valuable members of the sales force in numerous companies. These engineers must have interpersonal skills conducive to effective selling. Sales engineers bring many assets to their positions.

Engineers have the technical background to answer customer questions and concerns. They are trained to identify which products are right for the customer and how they can be applied. Sales engineers can also identify other applications or other products that might benefit the customer once they become familiar with their customer's needs.

In some sales forces, engineers are utilized because the customers are engineers themselves and have engineering-related questions. When airplane manufacturers market their aircraft to airlines, they send engineers. The airlines have engineers who have technical concerns overseeing the maintenance and operation of the aircraft. Sales engineers have the technical background to answer these questions.

As technology continues to advance, more and more products become technically sophisticated. This produces an ever-increasing demand for sales engineers.

Consulting

Consulting engineers are either self-employed or they work for a firm that does not provide goods or services directly to consumers. Such firms provide technical expertise to organizations that do. Many large companies do not have technical experts on staff in all areas of operation. Instead, they use consultants to handle issues in those technical areas.

For example, a manufacturing facility in which a cooling tower is used in part of the operation might have engineers who are well versed in the manufacturing processes but not in cooling tower design. The manufacturer would hire a consulting firm to design the cooling tower and related systems. Such a consultant might oversee the installation of such a system or simply provide a report with recommendations. After the system is in place or the report is delivered, the consultant would move on to another project with another company.

Consulting engineers also might be asked to evaluate the effectiveness of an organization. In such a situation, a team of consultants might work with a customer and provide suggestions and guidelines for improving the company's processes. These might be design methods, manufacturing operations, or even business practices. While some consulting firms provide only engineering-related expertise, other firms provide both engineering and business support and require the consulting engineers to work on business-related issues as well as technical issues.

Consulting engineers interact with a wide range of companies on a broad scope of projects, and come from all engineering disciplines. Often a consultant needs to be registered as a professional engineer in the state where he or she does business.

Management

In many instances, engineers work themselves into project management positions, and eventually into full-time management. National surveys show that more than half of all engineers will be involved in some type of management responsibilities—supervisory or administrative—before their career is over. Engineers are chosen for their technical ability, their problem-solving ability and their leadership skills.

Engineers may manage other engineers or support personnel, or they may rise to oversee the business aspects of a corporation. Often, prior to being promoted to this level of management, engineers acquire some business or management training. Some companies provide this training or offer incentives for employees to take management courses in the evening on their own time.

Other Fields

Some engineering graduates enter fields other than engineering, such as law, education, medicine, and business. Patent law is one area in which an engineering or science degree is almost essential. In patent law, lawyers research, write, and file patent applications. Patent lawyers must have the technical background to understand what an invention does so that they can properly describe the invention and legally protect it for the inventor.

Another area of law that has become popular for engineering graduates is corporate liability law. Corporations are faced with decisions every day over whether to introduce a new product, and must weigh potential risks. Lawyers with technical backgrounds have the ability to weigh technical as well as legal risks. Such lawyers are also used in litigation for liability issues. Understanding what an expert witness is saying in a suit can be a tremendous advantage in a courtroom and can enable a lawyer to effectively cross-examine the expert witness. Often, when a corporation is sued over product liability, the lawyers defending the corporation have technical backgrounds.

Engineers are involved in several aspects of education. The one students are most familiar with is engineering professors. College professors usually have their Ph.D.s. Engineers with master's degrees can teach at community colleges and in some engineering technology programs. Engineering graduates also teach in high schools, middle schools, and elementary schools. To do this full-time usually requires additional training in educational methods, but the engineering background is a great start. Thousands of engineers are involved in part-time educational projects where they go to classes as guest speakers to show students what "real engineers" do. The beginning of this chapter described one such engineer. The American Society for Engineering Education (ASEE) is a great resource if you are interested in engineering education.

Engineers also find careers in medicine, business, on Wall Street, and in many other professions. In modern society, with its rapid expansion of technology, the combination of problem-solving ability and technical knowledge makes engineers very valuable and extremely versatile.

EXAMPLE 2.3   A dean of a New York engineering school passed along a story about her faculty, which was concerned that most of that year's graduates had not accepted traditional engineering positions. The largest employers were firms that hire stockbrokers. The school's engineering graduates were prized for their ability to look at data and make rational conclusions from the data. In other words, they were very effective problem solvers. They also had the ability to model the data mathematically. This combination made the engineering graduates a perfect fit for Wall Street. This is not an isolated story. Many engineering graduates find themselves in careers unrelated to their engineering education. All engineers are trained problem solvers and in today's technically advanced society, such skills are highly regarded.

2.3 Engineering Majors

The following section is a partial listing of the various engineering disciplines in which engineering students can earn degrees. The list includes all of the most common majors. Some smaller institutions may offer programs that are not mentioned in this chapter. Be aware that some disciplines are referred to by different names at various institutions. Also, some programs may be offered as a subset of other disciplines. For instance, aeronautical and industrial engineering is sometimes combined with mechanical engineering. Environmental engineering might be offered as part of civil or chemical engineering.

These descriptions are not meant to be comprehensive. To describe a field completely, such as electrical engineering, would take an entire book by itself. These descriptions are meant to give you an overview of the fields and to answer some basic questions about the differences among the fields. It is meant to be a starting point. When selecting a major, a student might investigate several sources, including people actually working in the field.

An engineering student should keep in mind that the list of engineering fields is fluid. Areas such as aerospace, genetic, computer, and nuclear engineering did not exist 50 years ago. Yet men and women developed the technology to create these fields. In your lifetime, other new fields will be created. The objective is to gain the solid background and tools to handle future challenges. It is also important to find a field you enjoy.

EXAMPLE 2.4

"Show me the money!" As academic advisors, we see students regularly who are interested in engineering but not sure which discipline to pursue. Students are tempted to decide which field to enter by asking which offers the greatest salary. There are two issues a student should consider when looking at salaries of engineers.

First, all engineers make a good starting salary—well over the average American household income. However, a high salary would not make up for a job that is hated. The salary spread between the engineering disciplines is not large, varying only by about 10% from the average. The career you embark on after graduation from college will span some forty years—a long time to spend in a discipline you don't really enjoy.

Second, if you consider money a critical factor, you should consider earning potential, not starting salary. Earning potential in engineering is dependent, to a large extent, on job performance. Engineers who do well financially are those who excel professionally. Again, it is very rare for someone to excel in a field that he or she does not enjoy.

Aerospace Engineering

Aerospace engineering was previously referred to as aeronautical and astronautical engineering. Technically, aerospace engineering involves flight within the Earth's atmosphere and had its birth when two bicycle repairmen from Dayton, Ohio, made the first flight at Kitty Hawk, North Carolina, in 1903. Since that time, aerospace engineers have designed and produced aircraft that have broken the sound barrier, achieved near-orbit altitudes, and become the standard mode of transportation for long journeys. Aerospace engineering involves flight in space which began on October 4, 1957, when the Soviet Union launched Sputnik into orbit. The United States achieved one of its proudest moments in the field of aerospace on July 20, 1969, when Neil Armstrong set foot on the moon's surface—the first person ever to do so. In order to describe what the broad field of aerospace engineering has become, it can be separated into the categories of aerodynamics, propulsion, structures, controls, orbital mechanics, and life sciences.

Aerodynamics is the study of the flow of air over a streamlined surface or body. The aerodynamicist is concerned with the lift that is produced by a body such as the wing of an aircraft. Similarly, there is a resistance or drag force when a body moves through a fluid. An engineer looks to optimize lift and minimize drag. While mechanical, civil, and chemical engineers also study flow over bodies, the aerospace engineer is interested especially in high-speed air flows. When the speed of the aircraft approaches or exceeds the speed of sound, the modeling of the airflow becomes much more complex.

Air-breathing propulsion systems that power airplanes and helicopters include propellers, and gas turbine jets, as well as ramjets and scram-jets. Propulsion systems used to launch or operate spacecraft include liquid and solid rocket engines. **Propulsion engineers** are continually developing more efficient, quieter, and cleaner-burning conventional engines as well as new engine technologies. New engine concepts include wave rotors, electric propulsion, and nuclear-powered craft.

The structural support of an aircraft or spacecraft is critical. It must be both lightweight and durable. These conditions are often mutually exclusive and provide challenges for structural design engineers. **Structural engineers** utilize new alloys, composites, and other new materials to meet design requirements for more efficient and more maneuverable aircraft and spacecraft.

The control schemes for aircraft and spacecraft have evolved rapidly. The new commercial airplanes are completely digitally controlled. Aerospace engineers work with electrical engineers to design **control systems** and subsystems used to operate the craft. They must be able to understand the electrical and computational aspects of the control schemes as well as the physical systems that are being controlled.

Aerospace engineers are also interested in **orbital mechanics**. They calculate where to place a satellite to operate as a communication satellite or as a global positioning system (GPS). They might also determine how to use the gravity fields of the near-Earth planets to help propel a satellite to the outskirts of the solar system.

An aerospace engineer must be aware of human limitations and the effects of their craft on the human body. It would serve no useful purpose to design a fighter plane that was so maneuverable it incapacitated the pilot with a high G-force. Designing a plane and a system which keeps the pilot conscious is an obvious necessity. Understanding the physiological and psychological effects of lengthy exposure to weightlessness is important when designing a spacecraft or space station.

The American Institute of Aeronautics and Astronautics (AIAA) is one of the most prominent aerospace professional societies. It is composed of the following seven technical groups, which include 66 technical committees.

- Engineering and Technology Management
- Aircraft Technology Integration and Operations
- Propulsion and Energy
- Space and Missile Systems
- Aerospace Sciences
- Information and Logistics Systems
- Structures, Design, and Testing

Agricultural Engineering

Agricultural engineering traces its roots back thousands of years to the time when people began to examine ways to produce food and food products more efficiently. Today, the role of the agricultural engineer is critical to our ability to feed the ever-expanding population of the world. The production and processing of agricultural products is the primary concern of agricultural engineers. Areas within agricultural engineering include power machinery, food processing, soils and water, structures, electrical technologies, and bioengineering.

The mechanical equipment used on modern farms is highly sophisticated and specialized. **Harvesting equipment** not only removes the crop from the field but also begins to process it and provides information to the farmers on the quality of the harvest. This requires highly complicated mechanical and electrical systems that are designed and developed by the agricultural engineer.

Often, once the food is harvested it must be processed before it reaches the marketplace. Many different technologies are involved in the efficient and safe processing and delivery of food products. **Food process engineers** are concerned with providing healthier products to consumers who increasingly rely on processed food products. This often requires the agricultural engineer to develop new processes. Another modern-day concern is increased food safety. Agricultural engineers design and develop means by which food is produced free of contamination such as the irradiation techniques used to kill potentially harmful bacteria in food.

The effective management of **soil and water resources** is a key aspect of productive agriculture. Agricultural engineers design and develop means to address effective land use, proper drainage, and erosion control. This includes such activities as designing and implementing an irrigation system for crop production and designing a terracing system to prevent erosion in a hilly region.

Agricultural **structures** are used to house harvested crops, livestock, and their feed. Agricultural engineers design structures including barns, silos, dryers, and processing centers. They look for ways to minimize waste or losses, optimize yields, and protect the environment.

Electrical and information technology development is another area important to the agriculture community due to the fact that farms are typically located in isolated regions. Agricultural engineers design systems that meet the needs of the rural communities.

Bioengineering has rapidly evolved into a field with wide uses in health products as well as agricultural products. Agricultural engineers are working to harness these rapidly developing technologies to further improve the quality and quantity of the agricultural products necessary to continue to feed the world's population.

The American Society of Agricultural Engineers (ASAE) is one of the most prominent professional societies for agricultural engineers. It has eight technical divisions which address areas of agricultural engineering:

- Food Processing
- Information and Electrical Technologies
- Power and Machinery
- Structures and Environmental

- Soil and Water
- Forest
- Bioengineering
- Aqua Culture

Architectural Engineering

In ancient times, major building projects required a master builder. The master builder was responsible for the design and appearance of the structure, for selecting the appropriate materials, and for personally overseeing construction. With the advent of steel construction in the 19th century, it became necessary for master builders to consult with specialists in steel construction in order to complete their designs. As projects became more complex, other specialists were needed in such areas as mechanical and electrical systems. Modern architectural engineers facilitate the coordination between the creativity of the architect and the technical competencies of a variety of technology specialists.

Architectural engineers are well-grounded in the engineering fundamentals of structural, mechanical, and electrical systems, and have at least a foundational understanding of aesthetic design. They know how to design a building and how the various technical systems are interwoven within that design. Their strengths are both creative and pragmatic.

The National Society of Architectural Engineers, which is now the Architectural Engineering Institute, defined architectural engineering as:

> *". . . the profession in which a knowledge of mathematics and natural sciences, gained by study, experience, and practice, is applied with judgment to the development of ways to use, economically and safely, the materials and forces of nature in the engineering design and construction of buildings and their environmental systems."* [Belcher]

There are four main divisions within architectural engineering: structural; electrical and lighting; mechanical systems; and construction engineering and management.

Structural is primarily concerned with the integrity of the structure of buildings. Determining the integrity of a building's structure involves the analysis of loads and forces resulting from normal usage and operation as well as from earthquakes, wind forces, snow loads, or wave impacts. The structural engineer takes the information from the analysis, and designs the structural elements that will support the building while at the same time meeting the aesthetic and functional needs involved.

The area of **electrical and lighting** systems is concerned with the distribution of utilities and power throughout the building. This requires knowledge of electricity and power distribution as well as lighting concerns. Lighting restrictions may require energy-efficient lighting systems that are sufficient to meet the functional requirements related to use of the building. The architectural engineer also must insure that the lighting will complement the architectural designs of the building and the rooms within that structure.

Mechanical systems control the climate of a building, which includes cooling and heating the air in rooms as well as controlling humidity and air quality. Mechanical systems are also used to distribute water through plumbing systems. Other mechanical systems include transportation by way of elevators and escalators within a building. These systems must be integrated to complement the architectural features of the building.

The fourth area is **construction engineering and management**, which combine the technical requirements with the given financial and legal requirements to meet project deadlines. The architectural engineer is responsible for implementing the design in a way which assures the quality of the construction and meets the cost and schedule of the project. Modern computer tools and project management skills are implemented to manage such complex projects.

Other areas which are emerging for architectural engineers include energy management, computerized controls, and new building materials including plastics and composites and acoustics. Many programs have added explicit areas of emphasis in the application of alternative energy, energy efficiency, and sustainability into architectural engineering programs. Designing structures to reduce energy use, material consumption, and environmental impact in the design, construction, and operation phases of buildings is a growing area of interest and a topic of a great deal of research within the field.

Architectural engineers are employed by consulting firms, contractors, and government agencies. They may work on complex high-rise office buildings, factories, stadiums, research labs, or educational facilities. They also may work on renovating historic structures or developing affordable low-income housing. As the construction industry continues to grow and as projects continue to become more complex, the outlook for architectural engineers is bright.

There are less than 20 ABET accredited programs in architectural engineering in the United States. Some of these are four-year programs while others provide the combined engineering foundation and architectural insights in a five-year program.

The Architectural Engineering Institute (AEI) was formed in 1998 with the merger of the National Society of Architectural Engineers and the American Society of Civil Engineers Architectural Engineering Division. AEI was created to be the home for professionals in the building industry. It is organized into divisions, which include:

- Commercial Buildings
- Industrial Buildings
- Residential Buildings
- Institutional Buildings
- Military Facilities
- Program Management
- Mitigation of the Effects of Terrorism
- Building Systems
- Education
- Architectural Systems
- Fully Integrated and Automated Project Process
- Glass as an Engineered Material
- Sick/Healthy Buildings
- Designing for Facilities for the Aging

Reference: Belcher, M. C., *Magill's Survey of Science: Applied Science,* Salem Press, Pasadena, CA, 1993.

Biological Engineering

Biological engineering is one of the newer fields of engineering that spans the fields of biology and engineering. Traditionally, engineering has applied concepts from mathematics and the physical sciences to design, analyze, and manufacture inanimate products and processes to address human and environmental needs. Biological engineering, also called bioengineering, or biotechnological engineering, integrates the study of living systems into engineering, opening opportunities to adapt and even create living products. A search of the Accreditation Board of Engineering and Technology's website (www.abet.org, accessed November 2010) for accredited biological engineering shows a small but growing number of accredited programs. One example is at the Massachusetts Institute of Technology, whose website defines this new discipline as

> The goal of this biological engineering discipline is to advance fundamental understanding of how biological systems operate and to develop effective biology-based technologies for applications across a wide spectrum of societal needs including breakthroughs in diagnosis, treatment, and prevention of disease, in design of novel materials, devices, and processes, and in enhancing environmental health. web.mit.edu/be/index.shtml (accessed November 2010)

This definition shows the breadth and opportunities for this emerging field. Biological engineering is being practiced as separate programs as well as integrated with more traditional engineering majors at many schools. Some schools, for example, include biological engineering as a complement to their Agricultural Engineering Programs to differentiate the parts of their curriculum and research programs that address new opportunities in biological areas that have been opened via modern technology. Another example is the inclusion of bioengineering with biomedical engineering programs. Similar to other fields of engineering that have been grown out of the more traditional disciplines, Biological Engineering is evolving rapidly. At this point in the discipline's development, this means that the term "biological engineering" may mean slightly different things at different institutions. Students are encouraged to research individual programs to understand what this exciting and emerging field means at their respective institution.

Biomedical Engineering

Biomedical engineering is one of the newer fields of engineering, first recognized as a discipline in the 1940s. Its origins, however, date back to the first artificial limbs made of wood or other materials. Captain Ahab in "Moby Dick" probably didn't think of the person who made his wooden leg as a biomedical engineer, yet he was a predecessor for biomedical engineers who design modern prosthetics.

Biomedical engineering is a very broad field that overlaps with several other engineering disciplines. In some institutions, it may be a specialization within another discipline. Biomedical engineering applies the fundamentals of engineering to meet the needs of the medical community. Because of the wide range of skills needed in the fields of engineering and medicine, biomedical engineering often requires graduate work. The broad field of biomedical engineering encompasses the three basic categories of bioengineering, medical, and clinical.

Bioengineering is the application of engineering principles to biological systems. This can be seen in the production of food or in genetic manipulation to produce a disease- resistant strain of a plant or animal. Bioengineers work with geneticists to produce bioengineered products. New medical treatments are produced using genetically altered bacteria to produce human proteins needed to cure diseases. Bioengineers may work with geneticists in producing these new products in the mass quantities needed by consumers.

The medical aspect of biomedical engineering involves the design and development of devices to solve medical challenges. This includes designing mechanical devices such as a prosthesis which gives individuals the desired mobility. It also involves the development of the chemical processes necessary to make an artificial kidney function and the electrical challenges in designing a new pacemaker.

Medical engineers develop instrumentation for medical uses including non-intrusive surgical instruments. Much of the trauma of surgery results from the incisions made to gain access to the area of concern. Procedures allowing a surgeon to make a small incision, such as orthoscopic surgery, have greatly reduced risk and recovery time for patients.

Rehabilitation is another area in which biomedical engineers work. It involves designing and developing devices for the physically impaired. Such devices can expand the capabilities of impaired individuals, thereby improving their quality of life or shortening recovery times.

Clinical engineering involves the development of systems to serve hospitals and clinics. Such systems exhaust anesthetic gases from an operating room without impairing the surgical team. Air lines in ventilating systems must be decontaminated to prevent micro-organisms from spreading throughout the hospital. Rehabilitation centers must be designed to meet the needs of patients and staff.

Chemical Engineering

Chemical engineering differs from most of the other fields of engineering in its emphasis on chemistry and the chemical nature of products and processes. Chemical engineers take what chemists do in a laboratory and, applying fundamental engineering, chemistry, and physics principles, design and develop processes to mass-produce products for use in our society. These products include detergents, paints, plastics, fertilizers, petroleum products, food products, pharmaceuticals, electronic circuit boards, and many others.

The most common employment of chemical engineers is in the design, development, and operation of large-scale chemical production facilities. In this area, the design function involves the **design of the processes** needed to safely and reliably produce the final product. This may involve controlling and using chemical reactions, separation processes, or heat and mass transfer. While the chemist might develop a new compound in the laboratory, the chemical engineer would develop a new process to make the compound in a pilot plant, which is a small-scale version of a full-size production facility. An example would be the design of a process to produce a lower-saturated-fat product with the same nutritional value yet still affordable. Another example would be the development of a process to produce a higher-strength plastic used for automobile bumpers.

With respect to **energy sources**, chemical engineers are involved in the development of processes to extract and refine crude oil and natural gas. Petroleum engineering grew out of chemical engineering and is described in a separate section. Chemical engineers are also involved in alternative fuel development and production.

The processing of petroleum into plastics by chemical engineers has created a host of consumer products that are used every day. Chemical engineers are involved in adapting these products to meet new needs. GE Plastics, for instance, operates a research house in Massachusetts made entirely from plastic. It is used as a research facility to develop new building materials that are cheaper and more beneficial to the consumer.

Many of the chemicals and their byproducts used in industry can be dangerous to people and/or the environment. Chemical engineers must develop processes that minimize harmful waste. They work with both new and traditional processes to treat hazardous byproducts and reduce harmful emissions.

Chemical engineers are also very active in the bio-products arena. This includes the **pharmaceutical industry**, where chemical engineers design processes to manufacture new lines of affordable drugs. Geneticists have developed the means to artificially produce human proteins that can be used successfully to treat many diseases. They use genetically altered bacteria to produce these proteins. It is the job of the chemical engineer to take the process from the laboratory and apply it to a larger-scale production process to produce the quantities needed by society.

Chemical engineers are also involved in **bio-processes** such as dialysis, where chemical engineers work along with other biomedical engineers to develop new ways to treat people. Chemical engineers might be involved in the development of an artificial kidney. They could also be involved in new ways to deliver medicines, such as through skin implants or patches.

Chemical engineers have become very active in the production of **circuit boards**, such as those used in computers. To manufacture the very small circuits required in modern electronic devices, material must be removed and deposited very precisely along the path of the circuit. This is done using chemical techniques developed by chemical engineers. With the demand for smaller and faster electronics, challenges continue for chemical engineers to develop the processes to make them possible.

Chemical engineers are also involved in the **modeling of systems**. Computer models of manufacturing processes are made so that modifications can be examined without having to build expensive new facilities. Managing large processes involving facilities which could be spread around the globe requires sophisticated scheduling capabilities. Chemical engineers who developed these capabilities for chemical plants have found that other large industries with scheduling challenges, such as airlines, can utilize these same tools.

The American Institute of Chemical Engineering (AIChE) is one of the most prominent professional organizations for chemical engineers. It is organized into 13 technical divisions that represent the diverse areas of the chemical engineering field:

- Catalysis and Reaction Engineering
- Computing and Systems Technology
- Engineering and Construction Contracting
- Environmental
- Food, Pharmaceutical, and Bioengineering
- Forest Products
- Fuels and Petrochemicals
- Heat Transfer and Energy Conversion
- Management
- Materials Engineering and Sciences
- Nuclear Engineering
- Safety and Health
- Separations

Civil Engineering

Ancient examples of early **civil engineering** can be seen in the pyramids of Egypt, the Roman roads, bridges and aqueducts of Europe, and the Great Wall in China. These all were designed and built under the direction of the predecessors to today's civil engineers. Modern civil engineering continues to face the challenges of meeting the needs of society. The broad field of civil engineering includes these categories: structural, environmental, transportation, water resources, surveying, urban planning, and construction engineering.

A humorous aside perhaps illustrates one aspect of civil engineering. . . . An Air Force general with an engineering background was asked what the difference is between civil and aerospace engineers. "That's easy," he responded. "The aerospace engineers build modern, state-of-the-art weapon systems and the civil engineers build the targets."

Structural engineers are the most common type of civil engineer. They are primarily concerned with the integrity of the structure of buildings, bridges, dams, and highways. Structural engineers evaluate the loads and forces to which a structure will be subjected. They analyze the structural design in regard to earthquakes, wind forces, snow loads, or wave impacts, depending on the area in which the building will be constructed.

A related field of structural engineering is **architectural engineering**, which is concerned with the form, function and appearance of a structure. The architectural engineer works alongside the architect to ensure the structural integrity of a building. The architectural engineer combines analytical ability with the concerns of the architect.

Civil engineers in the **environmental** area may be concerned with the proper disposal of wastes—residential and industrial. They may design and adapt landfills and waste treatment facilities to meet community needs. Industrial waste often presents a greater challenge because it may contain heavy metals or other toxins which require special disposal procedures. Environmental engineering has come to encompass much more and is detailed in a later section of this chapter.

Transportation engineers are concerned with the design and construction of highways, railroads, and mass transit systems. They are also involved in the optimization and operation of the systems. An example of this is **traffic engineering**. Civil engineers develop the tools to measure the need for traffic control devices such as signal lights and to optimize these devices to allow proper traffic flow. This can become very complex in cities where the road systems were designed and built long ago but the areas around those roads have changed significantly with time.

Civil engineers also work with water resources as they construct and maintain dams, aqueducts, canals, and reservoirs. **Water resource engineers** are charged with providing safe and reliable supplies of water for communities. This includes the design and operation of purification systems and testing procedures. As communities continue to grow, so do the challenges for civil engineers to produce safe and reliable water supplies.

Before any large construction project can be started, the construction site and surrounding area must be mapped or surveyed. **Surveyors** locate property lines and establish alignment and proper placement of engineering projects. Modern surveyors use satellite technology as well as aerial and terrestrial photogrammetry. They also rely on computer processing of photographic data.

A city is much more than just a collection of buildings and roads. It is a home and working place for people. As such, the needs of the inhabitants must be taken into account in the design of the city's infrastructure. **Urban planning engineers** are involved in this process. They incorporate the components of a city (buildings, roads, schools, airports, etc.) to meet the overall needs of the population. These needs include adequate housing, efficient transportation, and open spaces. The urban planning engineer is always looking toward the future to fix problems before they reach a critical stage.

Another civil engineering area is **construction engineering**. In some institutions, this may even be a separate program. Construction engineers are concerned with the management and operation of construction projects. They are also interested in the improvement of construction methods and materials. Construction engineers design and develop building techniques and building materials that are safer, more reliable, cost-effective, and environmentally friendly. There are many technical challenges that construction engineers face and will continue to face in the coming decades.

An example of one such construction challenge is the rebuilding of a city's infrastructure, such as its sewers. While the construction of a sewer may not seem glamorous or state-of-the-art, consider the difficulty of rebuilding a sewer under an existing city. A real problem facing many cities is that their infrastructure was built decades or centuries ago, but it has a finite life. As these infrastructures near the end of their expected lives, the construction engineer must refurbish or reconstruct these systems without totally disrupting the city.

The American Society of Civil Engineers (ASCE) is one of the most prominent professional organizations for civil engineers. It is organized into the following 16 technical divisions covering the breadth of civil engineering:

- Aerospace
- Air Transport
- Architectural Engineering
- Construction Division
- Energy
- Engineering Mechanics
- Environmental Engineering
- Geomatics
- Highway
- Materials Engineering
- Pipeline
- Urban Planning and Development
- Urban Transportation
- Water Resources Engineering
- Water Resources Planning and Management
- Waterways, Ports, Coastal, and Ocean Engineering

Computer Engineering

Much of what computer and electrical engineers do overlaps. Many **computer engineering** programs are part of electrical engineering or computer science programs. However, computer technology and development have progressed so rapidly that specialization is required in this field, and thus computer engineering is treated separately. Given the wide range of computer applications and their continued growth, computer engineering has a very exciting future.

Computer engineering is similar to computer science, yet distinct. Both fields are extensively involved with the design and development of software. The main difference is that the computer scientist focuses primarily on the software and its optimization. The computer engineer, by contrast, focuses primarily on computer hardware—the machine itself. Software written by the computer engineer is often designed to control or to interface more efficiently with the hardware of the computer and its components.

Computer engineers are involved in the design and development of operating systems, compilers, and other software that requires efficient interfacing with the components of the computer. They also work to improve computer performance by optimizing the software and hardware in applications such as computer graphics.

Computer engineers work on the design of computer architecture. Designing faster and more efficient computing systems is a tremendous challenge. As faster computers are developed, applications arise for even faster machines. The continual quest for faster and smaller microprocessors involves overcoming barriers introduced by the speed of light and by circuitry so small that the molecular properties of components become important.

Computer engineers develop and design electronics to interface with computers. These include modems, Ethernet connections, and other means of data transmission. Computer engineers created the devices that made the Internet possible.

In addition to having computers communicate with each other, the computer engineer is also interested in having computers work together. This may involve increasing the communication speed of computer networks or using tightly coupled multiple processors to improve the computing speed.

Security is becoming a bigger concern as more and more information is transferred using computers. Computer engineers are developing new means of commercial and personal security to protect the integrity of electronic communications.

Artificial intelligence, voice recognition systems, and touch screens are examples of other technologies with which computer engineers are involved. The computers of the future will look and operate much differently than they do now, and it is the computer engineer who will make this happen.

Electrical Engineering

Considering the wide range of electronic devices people use every day, it is not surprising that **electrical engineering** has become the most populated of the engineering disciplines. Electrical engineers have a wide range of career opportunities in almost all of the industrial sectors. In order to provide a brief discussion of such a broad field, we will divide electrical engineering into eight areas: computers, communications, circuits and solid state devices, control, instrumentation, signal processing, bioengineering, and power.

Engineers specializing in **computer** technology are in such high demand that numerous institutions offer a separate major for computer engineering. Please refer back to the earlier section which described computer engineering separately.

Electrical engineers are responsible for the explosion in **communication** technologies. Satellites provide nearly instantaneous global communication. Global positioning systems (GPS) allow anyone with the required handheld unit to pinpoint precisely where they are located anywhere in the world. Fiber optics and lasers are rapidly improving the reliability and speed with which information can be exchanged. Wireless communication allows people to communicate anywhere, with anyone. Future breakthroughs in this field will have a tremendous impact on how we live in tomorrow's society.

Electrical engineers also design and develop electronic **circuits**. Circuit design has changed rapidly with the advent of microelectronics. As circuits continue to shrink, new barriers such as molecular size emerge. As the limits of current technology are approached, there will be incentives to develop new and faster ways to accomplish the same tasks.

Almost all modern machines and systems are digitally controlled. **Digital controls** allow for safer and more efficient operation. Electronic systems monitor processes and make corrections faster and more effectively than human operators. This improves reliability, efficiency, and safety. The electrical engineer is involved in the design, development, and operation of these control systems. The engineer must determine the kind of control required, what parameters to monitor, the speed of the correction, and many other factors. Control systems are used in chemical plants, power plants, automotive engines, airplanes, and a variety of other applications.

For a control system to operate correctly, it must be able to measure the important parameters of whatever it is controlling. Doing so requires accurate **instrumentation**, another area for the electrical engineer. Electrical engineers who work in this area develop electrical devices to measure quantities such as pressure, temperature, flow rate, speed, heart rate, and blood pressure. Often, the electrical devices convert the measured quantity to an electrical signal that can be read by a control system or a computer. Instrumentation engineers also design systems to transmit the measured information to a recording device, using telemetry. For example, such systems are needed for transmitting a satellite's measurements to the recording computers back on earth as it orbits a distant planet.

Signal processing is another area where electrical engineers are needed. In many instances the electrical signals coming from instrumentation or other sources must be conditioned before the information can be used. Signals may need to be electronically filtered, amplified, or modified. An example is the active noise control system on a stethoscope which allows a paramedic to listen to a patient's heart while in a helicopter. The active noise control system can block out the sound of the helicopter so the paramedic can make a quick, accurate assessment of the patient. Signal processing also comes into play in areas such as voice recognition for computers.

Electrical engineers also work in biomedical, or **bioengineering,** applications, as described earlier. Electrical engineers work with medical personnel to design and develop devices used in the diagnosis and treatment of patients. Examples include non-intrusive techniques for detecting tumors through magnetic resonance imaging (MRI) or through computerized axial tomography (CAT) scans. Other examples include pacemakers, cardiac monitors, and controllable prosthetic devices.

The generation, transmission, and distribution of electric **power** is perhaps the most traditional aspect of electrical engineering. Electrical engineers work closely with mechanical engineers in the production of electrical power. They also oversee the distribution of power through electrical networks and must ensure reliable supplies of electricity to our communities. With modern society's dependence on electricity, interruptions in the flow of electricity can be catastrophic. Many of today's power-related challenges revolve around reliability and cost of delivery. Electrical engineers also work with materials engineers to incorporate superconductivity and other technology in more efficient power transmission.

The Institute of Electrical and Electronics Engineers (IEEE) is the largest and most prominent professional organization for electrical engineers. It is organized into 37 technical divisions, which indicates the breadth of the field of electrical engineering. The technical divisions include:

- Aerospace and Electronic Systems
- Antennas and Propagation
- Broadcast Technology
- Circuits and Systems
- Communications
- Components Packaging and Manufacturing Technology
- Computer
- Consumer Electronics
- Control Systems
- Dielectrics and Electrical Insulation
- Education
- Electromagnetic Compatibility
- Electron Devices
- Engineering in Medicine and Biology
- Engineering Management
- Instrumentation and Measurement
- Lasers and Electro-Optics
- Magnetics
- Microwave Theory and Techniques
- Neural Networks
- Nuclear and Plasma Sciences
- Oceanic Engineering
- Power Electronics
- Power Engineering
- Professional Communication
- Reliability
- Robotics and Automation
- Signal Processing
- Social Implications of Technology
- Solid-State Circuits
- Geoscience and Remote Sensing

- Industrial Electronics
- Industrial Applications
- Information Theory
- Systems, Man, and Cybernetics
- Ultrasonics, Ferroelectrics, and Frequency Control
- Vehicular Technology

Environmental Engineering

Environmental engineering is a field which has evolved to improve and protect the environment while maintaining the rapid pace of industrial activity. This challenging task has three parts to it: disposal, remediation, and prevention.

Disposal is similar to that covered under civil engineering. Environmental engineers are concerned with disposal and processing of both industrial and residential waste. Landfills and waste treatment facilities are designed for residential waste concerns. The heavy metals and other toxins found in industrial wastes require special disposal procedures. The environmental engineer develops the techniques to properly dispose of such waste.

Remediation involves the cleaning up of a contaminated site. Such a site may contain waste which was improperly disposed of, requiring the ground and/or water to be removed or decontaminated. The environmental engineer develops the means to remove the contamination and return the area to a usable form.

Prevention is an area that environmental engineers are becoming more involved in. Environmental engineers work with manufacturing engineers to design processes that reduce or eliminate harmful waste. One example is in the cleaning of machined parts. Coolant is sprayed on metal parts as they are cut to extend the life of the cutting tools. The oily fluid clings to the parts and has to be removed before the parts are assembled. An extremely toxic substance had been used in the past to clean the parts because there was not a suitable alternative. Environmental engineers discovered that oil from orange peels works just as well and is perfectly safe (even edible). Since this new degreasing fluid did not need to be disposed of in any special way, manufacturing costs were reduced, and the environment of the workers improved.

Prevention does not just mean avoiding disasters. It also means reducing the environmental impact or "footprint" of products and processes. Environmental engineers may be engaged in improving the sustainability of processes, including manufacturing and construction as well as the design and operation of new products. The environmental engineers study the environmental impact of the processes or products; identify alternatives; and work to implement these alternatives to reduce the impact on our environment. Reducing the environmental impact of the products we use every day is vital to protecting the global environment as the world's population continues to grow.

Environmental engineers must be well grounded in engineering fundamentals and current on environmental regulations. Within their companies, they are the experts on compliance with the ever-changing environmental laws. An environmental engineer must be able to understand the regulations and know how to apply them to the various processes they encounter.

Industrial Engineering

Industrial engineering is described by the Institute of Industrial Engineers as the design, improvement, and installation of integrated systems of people, material, and energy. Industrial engineering is an interdisciplinary field that involves the integration of technology, mathematical models, and management practices. Traditional industrial engineering is done on a factory floor. However, the skills of an industrial engineer are transferable to a host of other applications. As a result, industrial engineers find themselves working within a wide variety of industries. Four of the main areas of emphasis for industrial engineers are production, manufacturing, human factors, and operations research.

The **production** area includes functions such as plant layout, material handling, scheduling, and quality and reliability control. An industrial engineer would examine the entire process involved in making a product, and optimize it by reducing cost and production time, and by increasing quality and reliability. In addition to factory layout, industrial engineers apply their expertise in other ways. For example, with an amusement park an industrial engineer would analyze the flow of people through the park to reduce bottlenecks and provide a pleasant experience for the patrons.

Manufacturing differs from production in that it addresses the components of the production process. While production concerns are on a global scale, manufacturing concerns address the individual production station. The actual material processing, such as machining, is optimized by the industrial engineer.

The **human factors area** involves the placement of people into the production system. An industrial engineer in this area studies the interfaces between people and machines in the system. The machines may include production machinery, computers, or even office chairs and desks. The industrial engineer considers ergonomics in finding ways to improve the interfaces. He or she looks for ways to improve productivity while providing a safe environment for workers.

Operations research is concerned with the optimization of systems. This involves mathematically modeling systems to identify ways to improve them. Project management techniques such as critical path identification fall under operations research. Often, computer simulations are required to either model the system or to study the effects of changes to the system. These systems may be manufacturing systems or other organizations. The optimizing of sales territories for a pharmaceutical sales force provides a non-manufacturing example.

The Institute for Industrial Engineering (IIE) is one of the most prominent professional organizations for industrial engineers. It is organized into three societies, 10 technical divisions and 8 interest groups. The following technical divisions show the breadth of industrial engineering:

- Aerospace and Defense
- Energy, Environment, and Plant Engineering
- Engineering Economy
- Facilities Planning and Design
- Financial Services
- Logistics Transportation and Distribution
- Manufacturing
- Operations Research
- Quality Control and Engineering Reliability
- Utilities

Marine and Ocean Engineering

Nearly 80 percent of the Earth's surface is covered with water. Engineers concerned with the exploration of the oceans, the transportation of products over water, and the utilization of resources in the world's oceans, seas, and lakes are involved in **marine and ocean engineering**.

Marine engineers focus on the design, development, and operation of ships and boats. They work together with **naval architects** in this capacity. Naval architects are concerned with the overall design of the ship. They focus on the shape of the hull in order to provide the appropriate hydrodynamic characteristics. They are also concerned with the usefulness of the vessel for its intended purpose, and with the design of the ship's subsystems including ventilation, water, and sanitary systems to allow the crew to work efficiently.

The **marine engineer** is primarily concerned with the subsystems of the ship which allow the ship to serve its purpose. These include the propulsion, steering, and navigation systems. The marine engineer might analyze the ship for vibrations or stability in the water. The ship's electrical power distribution and air-conditioning fall under the responsibility of marine engineers. They also might be involved in the analysis and design of the cargo handling systems of the ship.

The responsibilities of an **ocean engineer** involve the design, development, and operation of vehicles and devices other than boats or ships. These include submersible vehicles used in the exploration of the oceans and in the obtaining of resources from the ocean depths. He or she might be involved in the design of underwater pipelines or cables, offshore drilling platforms, and offshore harbor facilities.

Ocean engineers also are involved with the interaction of the oceans and things with which oceans come in contact. They study wave action on beaches, docks, buoys, moorings, and harbors. Ocean engineers design ways to reduce erosion while protecting the marine environment. They study ways to protect and maintain marine areas which are critical to our food supply. Ocean engineers become involved with pollution control and treatment in the sea and alternative sources of energy from the ocean.

One of the professional societies to which these engineers may be involved is the Society of Naval Architects and Marine Engineers. The society is subdivided into the following nine technical and research committees:

- Hull Structure
- Hydrodynamics
- Ship's Machinery
- Ship Technical Operations
- Offshore

- Ship Production
- Ship Design
- Ship Repair and Conversion
- Small Craft

Materials Engineering

The origins of **materials engineering** can be traced to around 3000 B.C. when people began to produce bronze for use in creating superior hunting tools. Since that time, many new materials have been developed to meet the needs of society. Materials engineers develop these new materials and the processes to create them. The materials may be metals or non-metals: ceramics, plastics, and composites. Materials engineers are generally concerned with four areas of materials: structure, properties, processes, and performance.

Materials engineers study the **structure** and composition of materials on a scale ranging from the microscopic to the macroscopic. They are interested in the molecular bonding and chemical composition of materials. The materials engineer is also concerned with the effect of grain size and structure on the material properties.

The **properties** in question might include strength, crack growth rates, hardness, and durability. Numerous technological advances are impeded by a lack of materials possessing the properties required by the design engineers. Materials engineers seek to develop materials to meet these demands.

A given material may have very different properties depending on how the material is **processed**. Steel is a good example. Cooling can affect its properties drastically.

Steel that is allowed to cool in air will have different properties than steel that is cooled through immersion in a liquid. The composition of a material also can affect its properties. Materials such as metallic alloys contain trace elements that must be evenly distributed throughout the alloy to achieve the desired properties. If the trace elements are not well distributed or form clumps in the metal, the material will have very different properties than the desired alloy. This could cause a part made with the alloy to fail prematurely. Materials engineers design processes and testing procedures to ensure that the material has the desired properties.

The materials engineer also works to ensure that a material meets the **performance** needs of its application by designing testing procedures to ensure that these requirements are met. Both destructive and nondestructive testing techniques are used to serve this process.

Materials engineers develop new materials, improve traditional materials, and produce materials reliably and economically through synthesis and processing. Subspecialties of materials engineering, such as metallurgy and ceramics engineering, focus on classes of materials with similar properties.

Metallurgy involves the extraction of metals from naturally occurring ore for the development of alloys for engineering purposes. The metallurgical engineer is concerned with the composition, properties and performance of an alloy. Detailed investigation of a component failure often identifies design flaws in the system. The materials engineer can provide useful information regarding the condition of materials to the design engineer.

Ceramics is another area of materials engineering. In ceramic engineering, the naturally occurring materials of interest are clay and silicates, rather than an ore. These nonmetallic minerals are employed in the production of materials that are used in a wide range of applications, including the aerospace, computer, and electronic industries.

Other areas of materials engineering focus on polymers, plastics, and composites. **Composites** are composed of different kinds of materials which are synthesized to create a new material to meet some specific demands. Materials engineers are also involved in **biomedical** applications. Examples include the development of artificial tissue for skin grafts, or bone replacement materials for artificial joints.

One of the professional societies to which materials engineers may belong is the Minerals, Metals, and Materials Society. It is organized into the following five technical divisions:

- Electronic, Magnetic, and Photonic Materials
- Extraction and Processing
- Light Metals
- Materials Processing and Manufacturing
- Structural Materials

Mechanical Engineering

Mechanical engineering is one of the largest and broadest of the engineering disciplines. It is second only to electrical engineering in the number of engineers employed in the field. Fundamentally, mechanical engineering is concerned with machines and mechanical devices. Mechanical engineers are involved in the design, development, production, control, operation, and service of these devices. Mechanical engineering is composed of two main divisions: design and controls, and thermal sciences.

The **design** function is the most common function of mechanical engineering. It involves the detailed layout and assembly of the components of machines and devices. Mechanical engineers are concerned about the strength of parts and the stresses the parts will need to endure. They work closely with materials engineers to ensure that correct materials are chosen. Mechanical engineers must also ensure that the parts fit together by specifying detailed dimensions.

Another aspect of the design function is the design process itself. Mechanical engineers develop computational tools to aid the design engineer in optimizing a design. These tools speed the design process by automating time-intensive analyses.

Mechanical engineers are also interested in controlling the mechanical devices they design. **Control** of mechanical devices can involve mechanical or hydraulic controls. However, most modern control systems incorporate digital control schemes. The mechanical engineer models controls for the system and programs or designs the control algorithm.

The noise generated from mechanical devices is often a concern, so mechanical engineers are often involved in acoustics—the study of noise. The mechanical engineer works to minimize unwanted noise by identifying the source and designing ways to minimize it without sacrificing a machine's performance.

In the **thermal sciences**, mechanical engineers study the flow of fluids and the flow of energy between systems. Mechanical engineers deal with liquids, gases, and two-phase flows, which are combinations of liquids and non-liquids. Mechanical engineers might be concerned about how much power is required to supply water through piping systems in buildings. They might also be concerned with aerodynamic drag on automobiles.

The flow of energy due to a temperature difference is called **heat transfer**, another thermal science area in which mechanical engineers are involved. They predict and study the temperature of components in environments of operation. Modern personal computers have microprocessors that require cooling. Mechanical engineers design the cooling devices to allow the electronics to function properly.

Mechanical engineers design and develop **engines**. An engine is a device that produces mechanical work. Examples include internal combustion engines used in automobiles and gas turbine engines used in airplanes. Mechanical engineers are involved in the design of the mechanical components of the engines as well as the overall cycles and efficiencies of these devices.

Performance and efficiency are also concerns for mechanical engineers involved in the production of power in large **power generation systems.** Steam turbines, boilers, water pumps, and condensers are often used to generate electricity. Mechanical engineers design these mechanical components needed to produce the power that operates the generators. Mechanical engineers also are involved in alternative energy sources including solar and hydroelectric power, and alternative fuel engines and fuel cells.

Another area in the thermal sciences is **heating, ventilating, and air-conditioning** (HVAC). Mechanical engineers are involved in the climate control of buildings, which includes cooling and heating the air in buildings as well as controlling humidity. In doing so, they work closely with civil engineers in designing buildings to optimize the efficiency of these systems.

Mechanical engineers are involved in the **manufacturing processes** of many different industries. They design and develop the machines used in these processes, and develop more efficient processes. Often, this involves automating time-consuming or expensive procedures within a manufacturing process. Mechanical engineers also are involved in the development and use of **robotics** and other automated processes.

In the area of **biomedical engineering**, mechanical engineers help develop artificial limbs and joints that provide mobility to physically impaired individuals. They also develop mechanical devices used to aid in the diagnosis and treatment of patients.

The American Society of Mechanical Engineering (ASME) is one of the most prominent professional societies for mechanical engineers. It is divided into 35 technical divisions, indicating the diversity of this field. These divisions are:

- Advanced Energy Systems
- Aerospace Engineering
- Applied Mechanics
- Basic Engineering Technical Group
- Bioengineering
- Design Engineering
- Dynamic Systems and Control
- Electrical and Electronic Packaging
- FACT
- Fluids Engineering
- Fluids Power Systems and Technology Systems
- Heat Transfer
- Information Storage/Processing
- Internal Combustion Engine
- Gas Turbine
- Manufacturing Engineering
- Materials
- Materials Handling Engineering
- Noise Control and Acoustics
- Non-destructive Evaluation Engineering
- Nuclear Engineering
- Ocean Engineering
- Offshore Mechanics / Arctic Engineering
- Petroleum
- Plant Engineering and Maintenance
- Power
- Pressure Vessels and Piping
- Process Industries
- Rail Transportation
- Safety Engineering and Risk Analysis

- Solar Energy
- Solid Waste Processing
- Technology and Society
- Textile Engineering
- Tribology

Mining Engineering

Modern society requires a vast amount of products made from raw materials such as minerals. The continued production of these raw materials helps to keep society functioning. **Mining engineers** are responsible for maintaining the flow of these raw materials by discovering, removing, and processing minerals into the products society requires.

Discovering the ore involves exploration in conjunction with geologists and geophysicists. The engineers combine the utilization of seismic, satellite, and other technological data, utilizing a knowledge of rocks and soils. The exploration may focus on land areas, the ocean floor, or even below the ocean floor. In the future, mining engineers may also explore asteroids, which are rich in mineral deposits.

Once mineral deposits are identified, they may be **removed**. One way minerals are removed is by way of mining tunnels. The engineers design and maintain the tunnels and the required support systems including ventilation and drainage. Other times, minerals are removed from open pit mines. Again, the engineers analyze the removal site and design the procedure for removing the material. The engineer also develops a plan for returning the site to a natural state. Mining engineers use boring, tunneling, and blasting techniques to create a mine. Regardless of the removal technique, the environmental impact of the mining operation is taken into account and minimized.

The mining engineer is also involved in the **processing** of the raw minerals into usable forms. Purifying and separating minerals involves chemical and mechanical processes. While mining engineers may not be involved in producing a finished product that consumers recognize, they must understand the form their customers can use and design processes to transform the raw materials into usable forms.

Mining engineers also become involved in the **design** of the specialized equipment required for use in the mining industry. The design of automated equipment capable of performing the most dangerous mining jobs helps to increase safety and productivity of a mining operation. Since mines typically are established in remote areas, mining engineers are involved in the **transportation** of minerals to the processing facility.

The expertise of mining engineers is not used exclusively by the mining industry. The same boring technology used in developing mines is used to create subway systems and railroad tunnels, such as the one under the English Channel.

Nuclear Engineering

Nuclear engineers are concerned primarily with the use and control of energy from nuclear sources. This involves electricity production, propulsion systems, waste disposal, and radiation applications.

The production of **electricity** from nuclear energy is one of the most visible applications of nuclear engineering. Nuclear engineers focus on the design, development, and operation of nuclear power facilities. This involves using current fission technology as well as the development of fusion, which would allow sea water to be used as fuel. Nuclear energy offers an environmentally friendly alternative to fossil fuels. A current barrier to production of nuclear facilities is the high cost of construction. This barrier provides a challenge for design engineers to overcome. Research is currently being performed on the viability of smaller, more efficient nuclear reactors.

Nuclear power is also used in **propulsion systems**. It provides a power source for ships and submarines, allowing them to go years without refueling. It is also used as a power source for satellites. Nuclear-powered engines are being examined as an alternative to conventional fossil-fueled engines, making interplanetary travel possible.

One of the main drawbacks to nuclear power is the production of **radioactive waste**. This also creates opportunities for nuclear engineers to develop safe and reliable means to dispose of spent fuel. Nuclear engineers develop ways to reprocess the waste into less hazardous forms.

Another area in which nuclear engineers are involved is the use of **radiation** for medical or agricultural purposes. Radiation therapy has proven effective in treating cancers, and radioactive isotopes are also used in diagnosing diseases. Irradiating foods can eliminate harmful bacteria and help ensure a safer food supply.

Due to the complex nature of nuclear reactions, nuclear engineers are at the forefront of advanced computing methods. High-performance computing techniques, such as parallel processing, constitute research areas vital to nuclear engineering.

The American Nuclear Society (ANS) is one of the professional societies to which nuclear engineers belong. It is divided into these 16 technical divisions:

- Biology and Medicine
- Decommissioning, Decontamination, and Reutilization
- Education and Training
- Environmental Sciences
- Fuel Cycle and Waste Management
- Fusion Energy
- Human Factors
- Isotopes and Radiation
- Materials Science and Technology

- Mathematics and Computations
- Nuclear Criticality Safety
- Nuclear Operations
- Nuclear Installations Safety
- Power
- Radiation Protection and Shielding
- Reactor Physics

Petroleum Engineering

Petroleum and petroleum products are essential components in today's society. **Petroleum engineers** maintain the flow of petroleum in a safe and reliable manner. They are involved in the exploration for crude oil deposits, the removal of oil, and the transporting and refining of oil.

Petroleum engineers work with geologists and geophysicists to identify potential oil and gas reserves. They combine satellite information, seismic techniques, and geological information to **locate deposits of gas or oil**. Once a deposit has been identified, it can be removed. The petroleum engineer designs, develops, and operates the needed drilling equipment and facilities. Such facilities may be located on land or on offshore platforms. The engineer is interested in removing the oil or gas in a safe and reliable manner—safe for the people involved as well as for the environment. **Removal of oil** is done in stages, with the first stage being the easiest and using conventional means. Oil deposits are often located in sand. A significant amount of oil remains coating the sand after the initial oil removal. Recovery of this additional reserve requires the use of secondary and tertiary extraction techniques utilizing water, steam, or chemical means.

Transporting the oil or gas to a processing facility is another challenge for the petroleum engineer. At times this requires the design of a heated pipeline such as the one in Alaska to carry oil hundreds of miles over frozen tundra. In other instances this requires transporting oil in double-hulled tankers from an offshore platform near a wildlife refuge. Such situations necessitate extra precautions to ensure that the wildlife is not endangered.

Once the oil or gas arrives at the processing facility, it must be **refined** into usable products. The petroleum engineer designs, develops, and operates the equipment to chemically process the gas or oil into such end products. Petroleum is made into various grades of gasoline, diesel fuel, aircraft fuel, home heating oil, motor oils, and a host of consumer products from lubricants to plastics.

Bioengineering

Bioengineering is the application of biotechnology. It is also referred to as biological engineering or biosystems engineering, and involves the application of engineering principles to living things. The U.S. National Institute of Health defines bioengineering as integrating "physical, chemical, or mathematical sciences and engineering principles for the study of biology, medicine, behavior, or health."

Bioengineers can work in a variety of fields, including medicine and medical device development, drug manufacturing, agriculture, and environmental science. They can work in industry, hospitals, academic institutions, and government agencies, among others. Biomechanical engineers, for instance, help design equipment for surgeons to perform arthroscopic procedures. Other bioengineers work to design better exercise equipment or therapeutic devices. Some work with nanotechnology to repair cell damage or control gene function. Still others build artificial limbs or organs.

Green Engineering

Green engineering is a general term used to refer to the application of engineering principles in an effort to conserve natural resources or to minimize adverse impact on the environment. Efficiency, renewability, and durability are some of the key aspects of green design.

Examples of green engineering include efforts by architects and engineers to develop structures that consume a minimum of energy, both in their construction and in their ongoing operation. This may involve the use of solar energy or other design elements. LEED (Leadership in Energy and Environmental Design) Certification pertains to standards for construction related to energy conservation and can involve criteria related to material selection, determination of location of manufacture, water use efficiency, indoor air quality, and even landscaping and irrigation.

Green engineering principles are also applied to automobile manufacture, including hybrid vehicles and flex fuels. Other careers involve solar energy generation, thermal engineering, and providing clean water supplies.

Other Fields

The most common engineering majors have been described in this chapter. However, there are other specialized engineering programs at some institutions. Here is a partial listing of some of these other programs:

- Automotive Engineering
- Acoustical Engineering
- Applied Mathematics
- Engineering Science
- Engineering Management

- Excavation Engineering
- Fire Engineering
- Forest Engineering
- General Engineering
- Genetic Engineering
- Geological Engineering
- Inventive Design
- Manufacturing Engineering
- Packaging Engineering
- Pharmaceutical Engineering
- Plastics Engineering
- Power Engineering
- Systems Engineering
- Theatre Engineering
- Transportation Engineering
- Welding Engineering

2.4   **Emerging Fields**

The fields of engineering have been, and will continue to be, dynamic. As new technologies emerge, new definitions are needed to classify disciplines. The boundaries will continue to shift and new areas will emerge. The explosion of technological advances means that there is a good chance you will work in a field that is not currently defined. Technological advances are also blurring the traditional delineations between fields. Areas not traditionally linked are coming together and providing cross-disciplinary opportunities for engineers. A good example of this is in the development of smart buildings which sense the onset of an earthquake, adapt their structures to survive the shaking, and then return to normal status afterward. Research is being conducted on these technologies that bridge computer engineering with structural (civil) engineering to produce such adaptable buildings.

The incredible breakthroughs in biology have opened many new possibilities and will continue to impact most of the fields of engineering. Historically, biology has not been as integrated with engineering as has physics and chemistry, but that is rapidly changing and will have enormous impact on the future of engineering. The ability to modify genetic codes has implications in a wide range of engineering applications including the production of pharmaceuticals that are customized for individual patients, alternative energy sources, and environmental reclamation.

Nanotechnology is an area that is receiving a great deal of attention and resources, and is blurring the boundaries of the fields of engineering. Nanotechnology is an emerging field in which new materials and tiny structures are built atom-by-atom, or molecule-by-molecule, instead of the more conventional approach of sculpting parts from pre-existing materials. The possibilities for nano-applications include:

- the creation of entirely new materials with superior strength, electrical conductivity, resistance to heat, and other properties

- microscopic machines for a variety of uses, including probes that could aid diagnostics and repair
- a new class of ultra-small, super-powerful computers and other electronic devices, including spacecraft
- a technology in which biology and electronics are merged, creating "gene chips" that instantly detect food-borne contamination, dangerous substances in the blood, or chemical warfare agents in the air
- the development of "molecular electronics" and devices that "self assemble," similar to the growth of complex organic structures in living organisms

Nanotechnology requires specialized laboratory and production facilities that provide further challenges and opportunities for future engineers.

The explosion of information technology with the Internet and wireless communication has produced a melding of disciplines to form new fields in information science and technology. Information management and transfer is an important and emerging issue in all disciplines of engineering and has opened opportunities for engineers who want to bridge the gaps between the traditional fields and information technologies.

Sustainability is another important concept that spans many engineer disciplines. Sustainable development is a term that was defined by a report to the United Nations, known as the Brundtland Report, in 1987 as: "development that meets the needs of the present without compromising the ability of future generations to meet their own needs." (*Our Common Future, 1987*). Designing products and processes that contribute to sustainable development is important for all engineering disciplines and is becoming a larger part of undergraduate and graduate education as well as professional development programs. As the global population continues to rise, so too do the pressures on the global environment. Sustainability will have to play a large and significant role in product development, manufacturing, and construction if we are to meet these challenges.

Undoubtedly, more new fields will open and be discovered as technology continues to advance. As the boundaries of the genetic code and molecular-level device are crossed, new frontiers will open. As an engineer, you will have the exciting opportunity to be part of the discovery and definition of these emerging fields which will have tremendous impact on society's future.

2.5 **Closing Thoughts**

The information presented in this chapter is meant to provide a starting point on the road to choosing a career. There may have been aspects of one or more of these engineering fields that appealed to you, and that's great. The goal, however, is not to persuade you that engineering is for everyone; it is not. The goal is to provide information to help you decide if engineering would be an enjoyable career for *you*.

As a student, it is important to choose a career path that will be both enjoyable and rewarding. For many, an engineering degree is a gateway to just such a career. Each person has a unique set of talents, abilities, and gifts that are well matched for a particular career.

In general, people find more rewarding careers in occupations where their gifts and talents are well used. Does engineering match your talents, abilities, and interests?

Right now, choosing a career may seem overwhelming. But at this point, you don't have to. What you are embarking on is an education that will provide the base for such decisions. Think about where a degree in engineering could lead. One of the exciting aspects of an engineering education is that it opens up a wide range of jobs after leaving college. Most students will have several different careers before they retire, and the important objective in college is obtaining a solid background that will allow you to move into areas that you enjoy later in life.

As you try to make the right choice for you, seek out additional information from faculty, career centers, professional societies, placement services, and industrial representatives. Ask a lot of questions. Consider what you would enjoy studying for four years in college. What kind of entry level job would you be able to get with a specific degree? What doors would such a degree open for you later in life? Remember, your decision is unique to you. No one can make it for you.

REFERENCES

Burghardt, M.D., *Introduction to the Engineering Profession*, 2nd Edition, Harper Collins College Publishers, 1995.

Garcia, J., *Majoring {Engineering}*, The Noonday Press, New York, New York, 1995.

Grace, R., and J. Daniels, *Guide to 150 Popular College Majors*, College Entrance Examination Board, New York, New York, 1992, pp. 175–178.

Irwin, J.D., *On Becoming an Engineer*, IEEE Press, New York, 1997.

Kemper, J.D., *Engineers and Their Profession*, 4th Edition, Oxford University Press, New York, 1990.

Landis, R., *Studying Engineering, A Road Map to a Rewarding Career*, Discovery Press, Burbank, California, 1995.

Smith, R.J., B.R. Butler, W.K. LeBold, *Engineering as a Career*, 4th Edition, McGraw-Hill Book Company, New York, 1983.

World Commission on Environment and Development, *Our Common Future*, Oxford University Press, 1987.

Wright, P.H., *Introduction to Engineering*, 2nd Edition, John Wiley and Sons, Inc., New York, 1994.

EXERCISES AND ACTIVITIES

2.1   Contact a practicing engineer in a field of engineering that interests you. Write a brief report on his or her activities and compare them to the description of that field of engineering as described in this chapter.

2.2   For a field of engineering that interests you, make a list of potential employers that hire graduates from your campus, and list the cities in which they are located.

2.3   Visit a job fair on your campus and briefly interview a company representative. Prepare a brief report on what that company does, what the engineer you spoke to does, and the type of engineers they are looking to hire.

2.4 Make a list of companies that hire co-op and/or intern students from your campus. Write a brief report on what these companies are looking for and their locations.

2.5 Select a company that employs engineers in a discipline that interests you, and visit their web page. (You can search for them using Yahoo or Alta Vista, or you can call them and ask for their Web address.) Prepare a brief report on what the company does, hiring prospects, engineering jobs in that organization, and where the company is located.

2.6 Contact a person with an engineering degree who is not currently employed in a traditional engineering capacity. Write a one-page paper on how that person uses his or her engineering background in their job.

2.7 Write a one-page paper on an engineering field that will likely emerge during your lifetime (a field that does not currently exist). Consider what background an engineering student should obtain in preparation for this emerging field.

2.8 Draft a sample letter requesting a co-op or intern position.

2.9 Identify a modern technological problem. Write a brief paper on the role of engineers and technology in solving this problem.

2.10 Select an engineering discipline that interests you and a particular job function within that discipline. Write a brief paper contrasting the different experiences an engineer in this discipline would encounter in each of three different industries.

2.11 Make a list of your own strengths and talents. Write a brief report on how these strengths are well matched with a specific engineering discipline.

2.12 Pick an engineering discipline that interests you. List and briefly describe the technical and design electives available in that discipline for undergraduates.

2.13 Select a consumer product you are familiar with (a stereo, clock radio, automobile, food product, etc.). List and briefly describe the role of all the engineering disciplines involved in producing the product.

2.14 Select an engineering discipline that interests you. Write a brief paper on how the global marketplace has altered this discipline.

2.15 Write a brief paper listing two similarities and two differences for each of the following engineering functions:

a) research and development   d) manufacturing and operations

b) development and design   e) sales and customer support

c) design and manufacturing   f) management and consulting

2.16 Find out how many engineering programs your school offers. How many students graduate each year in each discipline?

2.17   Which of the job functions described in this chapter is most appealing to you? Write a brief paper discussing why it is appealing.

2.18   Write a paper about an engineer who made a significant technical contribution to society.

2.19   Report on the requirements for becoming a registered professional engineer in your state. Also report on how registration would be beneficial to your engineering career.

2.20   Make a list of general responsibilities and obligations you would have as an engineer.

2.21   Write a brief paper on the importance of ethical conduct as an engineer.

2.22   Select one of the industrial sectors within engineering and list five companies that do business in that sector. Briefly describe job opportunities in each.

2.23   Prepare a report on how the following items work and what engineering disciplines are involved in their design and manufacture:

   a) CD player   d) Dialysis machine
   b) CAT scan machine   e) Flat TV screen
   c) Computer disk drive

2.24   Answer the following questions and look for themes or commonality in your answers. Comment on how engineering might fit into these themes.

   a) If I could snap my fingers and do whatever I wanted, knowing that I wouldn't fail, what would I do?
   b) At the end of my life, I'd love to be able to look back and know that I had done something about _____.
   c) What would my friends say I'm really interested in or passionate about?
   d) What conversation could keep me talking late into the night?
   e) What were the top five positive experiences I've had in my life, and why were they meaningful to me?

2.25   Write a short paper describing your dream job, regardless of pay or geographical location.

2.26   Write a short paper describing how engineering might fit into your answer to 2.25.

2.27   List the characteristics a job must have for you to be excited to go to work every morning. How does engineering fit with those characteristics?

2.28   List the top ten reasons why a student should study engineering.

2.29   List the ten inappropriate reasons for a student to choose to study engineering.

2.30   Select an industrial sector and describe what you suppose a typical day is like for:

a) Sales engineer   d) Manufacturing engineer

b) Research engineer   e) Design engineer

c) Test engineer

2.31   Make a list of five non-engineering careers an engineering graduate could have and describe each one briefly.

2.32   Identify one of the student branches of an engineering professional society on your campus and prepare a report on the benefits of involvement with that organization for an engineering student.

2.33   Write a letter to a 9th-grade class explaining the exciting opportunities in your chosen major within engineering. Include a short discussion of the classes they should be taking to be successful in that same major.

2.34   Write a letter to a 9th-grade class describing the opportunities a bachelor's degree in engineering provides in today's society.

2.35   Write a letter to your parents detailing why you are going to major in the field you have chosen.

2.36   Write a one-page paper describing how your chosen field of engineering will be different in 25 years.

2.37   Select one field of engineering and write a one-page paper on how the advances in biology have influenced that field.

2.38   Select one field of engineering and write a one-page paper on how the advances in nanotechnology have influenced that field.

2.39   Select one field of engineering and write a one-page paper on how the advances in information technology have influenced that field.

2.40   Select two fields of engineering and describe problems that span these two disciplines.

2.41   Write a brief paper on an emerging area within engineering. Relate the area to your chosen engineering major.

2.42   Research the current spending priorities of the U.S. government in the area of technical research. How will these priorities impact your chosen major?

2.43   For each grouping of engineering disciplines, describe applications or problems that span the disciplines.

   a. Aerospace, Materials, and Civil

   b. Mechanical, Agricultural, and Computer

   c. Biological and Environmental

   d. Industrial, Chemical, and Electrical

   e. Biomedical and Nuclear

   f. Agricultural and Aerospace

   g. Materials and Biomedical

   h. Civil and Computer

2.44   Select one professional organization and find out how they handle new and emerging technologies within their society (where do they put them and where can people working in emerging areas find colleagues?).

2.45   Interview a practicing engineering and a faculty member from the same discipline about their field. Compare and contrast their views of the discipline.

2.46   Select two engineering majors and compare and contrast the opportunities available between the two.

2.47   Identify how your skills as an engineer can be used within your local community— either as full time work or as a volunteer. Share your findings with the class by preparing a short oral presentation.

CHAPTER 3

A Statistical Profile of the Engineering Profession

3.1 Statistical Overview

Many students who choose a major in engineering know little about the profession or the individuals who work in the field. This chapter will answer such questions as: How many people study engineering? What are their majors? What is the job market for engineers like? How much do engineers earn? How many women and minorities are studying engineering? How many practicing engineers are there in the U.S.?

The picture of the engineering profession presented in this chapter will assist you in better understanding the field. This information has been gathered from a variety of surveys and reports that are published on a regular basis by various organizations. As with many types of data, some are very current and others not as current. We have endeavored to provide commentary that explains the information presented.

3.2 College Enrollment Trends of Engineering Students

As observed in Figure 3.1, the number of first-year students enrolled in engineering programs has fluctuated greatly during the period from 1955 to 2011, according to the 2011 Engineering Workforce Commission's annual survey of enrollment patterns. During the 1950s and 1960s the number of first-year students pursuing bachelor's degrees in engineering programs ranged between 60,000 and 80,000. During the 1970s there was a significant drop in student interest, and freshman enrollment fell to a low of 43,000. This was followed by the largest long-term climb in student enrollment, which eventually peaked in the early 1980s at about 118,000 students. From that point until 1996 there was a steady decline in student enrollment, with a low of about 85,000 first-year students. This was followed in 1997 to 2002 with a modest increase in student enrollment each year, which was projected to last for several years. From 2003–2006 there was a decline in first-year students enrolling in engineering programs, even though total undergraduate engineering enrollment remained high. 2007 and 2008 showed a modest increase in first-year students to 111,000. However, beginning in 2009 to the present, there has been a significant increase: 2011 data show almost 122,000 first-year engineering students.

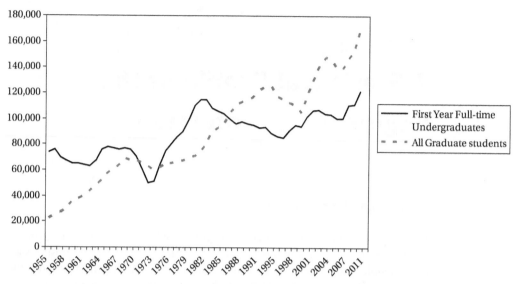

Figure 3.1 Engineering enrollments: selected indicators, 1955–2011

Sources: Through 1978, Table 23, "Engineering Enrollments and Degrees" in the Placement of Engineering And Technology Graduates, 1979 (New York: Engineering Manpower Commission); for 1979–1998, Engineering And Technology Enrollments, Fall 1979–1998, Engineering Workforce Commission of the American Association of Engineering Societies, 1979–2011, Reston, VA

Why has enrollment in engineering (and higher education in general) fluctuated so greatly over the last few decades? The primary factor influencing enrollment is the number of high school graduates in a given year. Of course, those numbers are directly correlated to the number of births 18 years previously. Given these facts and the information in Figure 3.2, it can be determined how the birth rates from 1960 to 2011 have influenced, and will influence, the enrollment of college-bound students. Likewise, Figure 3.3 provides data concerning the number of high school graduates from 1988 to 2021. This shows that the number of graduates started to climb in 1992 and increased in 1999 through 2005. There were modest increases in 2006 through 2009, and then projections show a slight decline followed by a rise in the number of graduates through 2021.

Figure 3.2 Births in the U.S., 1960–2011

Source: U.S. Department of Health and Human Services, Centers for Disease Control and Prevention National Center for Health Statistics, 2012

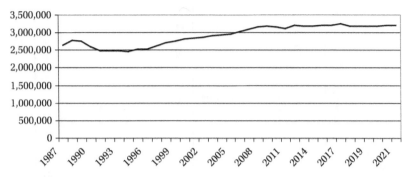

Figure 3.3 High school graduates, 1987–2021

Source: US Department of Education, National Center for Education Statistics, 2011. Projected data are for the years 2008–2021.

Obviously, there are other factors which influence a student's choice of a major, including the general economy and the relative popularity of various fields in any given year.

It also should be noted that the total number of students pursuing graduate engineering degrees increased significantly from the late 1970s to 1992. Since then, there was a relatively steady decline until 1997, when enrollment started increasing again, with 2004 numbers reaching an all-time high of over 145,000. 2005 and 2006 numbers show a decline to about 140,000, but 2008 figures rose to over 153,000 graduate students and the number has continued to climb to over 168,00 in 2011.

3.3 College Majors of Recent Engineering Students

Table 3.1 provides specific 2011 enrollment data for undergraduate and graduate students in several broad engineering fields. This information shows that over 511,000 students were majoring in one of the undergraduate engineering disciplines, with another 168,000 pursuing graduate engineering degrees. The largest field of study at all levels is the electrical and computer engineering disciplines, followed by mechanical and aerospace. It is interesting to note that over 102,000 students are in "other engineering disciplines" (which includes the many smaller, specialized fields of engineering) and over 18,000 are in pre-engineering programs.

3.4 Degrees in Engineering

When examining the number of degrees awarded to engineering graduates during the years 1983–2011 (Fig 3.4), there is a close correlation between enrollment and degrees. Since 1986, when over 78,000 engineering graduates were produced, to 1997, when approximately 65,000 degrees were awarded, there was a steady decline (with the exception of 1993) until 1995, and then a slight increase until 1997. However, there were declines in 1998 and 1999, and then a gradual increase between 2000 and 2006, to a total of 76,000. 2007, however, shows a slight decline in degrees awarded, followed by an increase in 2008 and a decline in 2009 and a significant increase in 2010 and 2011.

Table 3.1 Fall, 2011 Engineering Enrollments by Broad Disciplinary Groups

	All Engineering Disciplines	Electrical and Computer	Mechanical and Aerospace	Civil and Environmental	Chemical and Petroleum	Industrial, Management, & Manufacturing	All Other Disciplines	Pre-Engineering
Full-Time Undergrads								
First Year	121,766	26,375	26,194	12,366	8,373	2,147	35,519	10,792
Second Year	102,622	25,154	25,582	12,301	8,803	3,064	22,131	5,587
Third Year	101,811	28,259	26,907	14,423	9,410	4,110	17,475	1,227
Fourth Year	138,107	38,449	36,602	22,227	13,364	6,154	21,041	270
Fifth Year	7,614	1,817	2,123	1,196	807	316	1,331	24
Full-Time Undergrad	471,920	120,054	117,408	62,513	40,757	15,791	97,497	17,900
Part-Time Undergrad	39,386	14,860	9,512	5,844	2,091	1,233	5,168	678
Total All Undergrad	511,306	134,914	126,920	68,357	42,848	17,024	102,665	18,578
Candidates for M.S.								
Full Time	63,183	24,857	10,597	8,605	2,453	5,586	11,085	0
Part Time	38,507	13,482	5,750	4,344	798	4,728	9,402	3
Total All M.S.	101,690	38,339	16,347	12,949	3,251	10,314	20,487	3
Candidates for Ph.D.								
Full Time	58,114	20,323	9,457	5,369	5,783	2,204	14,973	—
Part Time	8,644	3,404	1,306	946	459	916	1,613	—
Total All Ph.D.	66,758	23,732	10,763	6,315	6,242	3,120	19,586	—
Full-Time Grads	121,297	45,185	20,054	13,974	8,236	7,790	26,053	—
Part-Time Grads	47,151	16,886	7,056	5,290	1,257	5,644	11,015	—
Total All Grads	168,448	62,071	27,110	19,264	9,493	13,434	37,073	—

Source: Engineering Workforce Commission of the American Association of Engineering Societies

Figure 3.4 Engineering degrees, 1983–2011, by degree level
Source: 1983–2011: Engineering Workforce Commission of the American Association of Engineering Societies

Table 3.2 documents the number of bachelor's degrees awarded by field of study for the years 2002–2011. It is interesting to note that the greatest number were earned by students in mechanical engineering, which in 2006 surpassed electrical engineering and the closely related fields of computer engineering or computer science (which at many schools are in the same department, or are closely aligned with electrical engineering programs).

3.5 Job Placement Trends

The Collegiate Employment Research Institute at Michigan State University conducts an annual hiring survey of employers and is able to track the labor market for new college graduates. From 1994 to mid-2001, the job market for graduating engineers was excellent; experts agree that it was probably the hottest job market in over 30 years. Unfortunately, overall hiring contracted nearly 50% in 2002–2003, due in part to the aftershock of 9/11. Hiring trends slowly recovered each year from 2004–2008, but then the economic downturn of 2008 began to negatively influence employment opportunities for new college graduates, including engineers, as available positions became highly competitive. It was not until 2010–2011 when a gradual shift in hiring patterns began to emerge. Students were once again finding that employers were returning to college campuses, job prospects were slowly improving, and more graduates were finding employment.

The most recent survey of over 4,300 employers for 2012–2013 shows a slowly growing economy and a continued tight labor market for new college graduates. If this trend continues, opportunities for engineers will still be available, but they will require that

Table 3.2 Bachelor's Degree in Engineering, by Discipline, 2002–2011

Curriculum	2002	2003	2004	2005	2006	2007	2008	2009	2010	2011
Aerospace	1,773	2,024	2,298	2,395	2,681	2,763	2,897	2,989	3,150	3,286
Agricultural	322	330	305	302	343	351	355	383	418	427
Biomedical	1,679	1,962	2,360	2,638	3,028	3,055	3,478	3,787	3,951	4,293
Chemical	5,657	5,342	4,939	4,621	4,590	4,607	4,915	5,151	5,849	6,297
Civil	8,185	8,595	8,477	8,857	9,432	9,875	10,862	11,274	11,835	13,175
Computer	13,703	16,607	17,101	16,019	14,282	13,171	11,471	10,370	10,764	11,610
Electrical	13,031	14,177	14,495	14,742	14,329	13,783	13,113	11,978	11,968	12,005
Engr. Science	923	1,018	1,072	1,019	1,015	1,097	1,104	1,104	1,212	1,233
Industrial	3,252	3,313	3,225	3,220	3,079	3,150	3,064	3,077	3,293	3,423
Materials & Metallurgical	803	869	775	815	919	924	1,027	985	1,111	1,134
Mechanical	13,343	14,114	14,321	14,835	15,698	16,172	17,865	17,016	18,040	19,016
Mining	305	276	206	224	243	345	345	341	380	417
Nuclear	144	123	195	239	327	324	400	394	412	463
Petroleum	259	330	269	298	381	468	541	665	784	994
Other Engineering Disciplines	5,269	5,951	5,965	5,779	5,756	5,738	5,670	5,806	6,361	6,826
Total	68,648	75,031	76,003	76,003	76,103	75,823	77,107	75,320	79,528	84,599

Source: Engineering Workforce Commission of the American Association of Engineering Societies, 2011

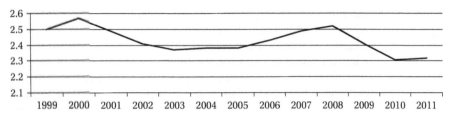

Figure 3.5 Numbers of employed engineers and other closely related occupations (in millions)

Source: Bureau of Labor Statistics, Occupational Employment Statistics, 2012

students devote increased efforts to finding the right opportunity. However, it should be noted that employment opportunities and starting salaries for engineers have traditionally surpassed those of most other majors on college campuses, and this trend continues.

The Bureau of Labor Statistics, in its 2012 report, shows that 2.3 million people work as engineers and other closely related professions. Figure 3.5 illustrates the recent trends in engineering employment over the last decade. These trends involve some modest fluctuation, perhaps due to corporate downsizing, and some engineering job losses. However, the overall picture is one of fairly stable employment.

Figure 3.6 illustrates the relatively low unemployment in the engineering profession. In 1994 the unemployment rate of engineers was about 3.5%. Since that time, there has been a steady decrease, until 2001 when the unemployment rate was around 1.7%.

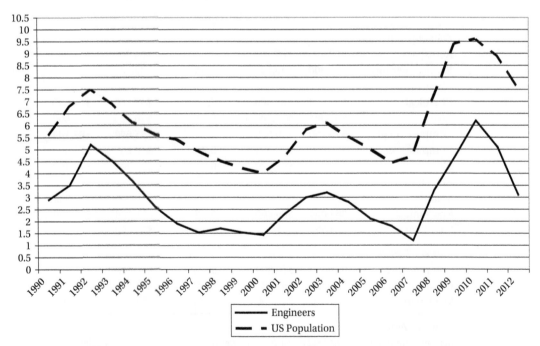

Figure 3.6 U.S. general population unemployment compared to engineering unemployment: recent trends, 1990–2012

Source: Bureau of Labor Statistics

Since 2001, engineering unemployment climbed over 3.0% in 2003, and then steadily dropped to 1.3% in 2007. However, 2008 numbers show engineering unemployment at 3.3%; the rate climbed to 6% in mid-2010. While the 2010 engineering unemployment rate was the highest since 1994, the unemployment rate of engineers continues to be well below that of the general population. 2011 and 2012 data show unemployment rates dropping for the entire workforce. The engineering figures show a similar decline.

3.6   **Salaries of Engineers**

Salaries of engineering graduates consistently have been among the highest for all college graduates over the past several decades. To some extent, this has been due to the shortage of engineering candidates to fill available jobs. However, even in "down" years and leaner economic periods, engineers tend to do better in the job market than do their counterparts from other majors.

As reflected in Table 3.3, recent data indicate that average starting salaries for 2013 graduates in selected engineering fields have varied significantly. Starting salary offers in many disciplines have increased while a few have dropped slightly. Petroleum

Table 3.3 Starting Salary Offers in Engineering Reported, April 2013

Curriculum	2013 Average	2012 Average	Percent Average Change in Starting Salaries	25th percentile	75th percentile
Bachelor's Degrees					
Aerospace	$64,400	$67,100	-4.0%	$54,000	$78,000
Bioengineering / Biomedical	$47,300	$43,200	9.5%	$36,000	$57,500
Chemical	$67,600	$64,300	5.1%	$55,100	$80,500
Civil	$57,300	$57,700	-0.7%	$46,500	$67,800
Computer Engr.	$71,700	$70,900	1.1%	$55,600	$82,100
Computer Sci.	$64,800	$61,600	5.2%	$50,800	$75,600
Electrical	$63,400	$59,700	6.2%	$51,400	$75,400
Industrial / Manufacturing	$56,300	*	N/A	$44,900	$67,300
Mechanical	$64,000	$60,600	5.6%	$51,100	$75,600
Petroleum	$93,500	*	N/A	$73,800	$109,700
Master's Degrees					
Computer Sci.	$74,200	$70,300	5.5%	$58,500	$84,400
Electrical	$67,200	$64,900	3.5%	$53,000	$78,800
Mechanical	$67,900	$66,800	1.6%	$53,900	$78,300

* Data Not Available

Reprinted from the April 2013 *Salary Survey*, with permission of the National Association of Colleges and Employers, copyright holder.

engineering B.S. graduates are at the top of the salary list. This trend may reflect the shortage of trained graduates in that field.

Table 3.4 provides starting salary data for 2013 graduates by engineering curriculum and some of the types of employers that hire students in those fields.

Table 3.4 Average 2013 Starting Salary by Curriculum and Some Common Types of Employers for Each Field

Curriculum / Type of Employer*	2013	Curriculum / Type of Employer*	2013
Bachelor's Degrees		▪ **Computer Science**	
▪ **Aerospace / Aeronautical Engineering**		Professional, Scientific, & Technical Services	$63,200
Manufacturing	$62,200	Information	$67,200
Professional, Scientific, & Technical Services	$64,100	Manufacturing	$68,700
		Finance and Insurance	$67,200
Federal, State & Local Government	$73,400	Management of Companies & Enterprises	$66,300
▪ **Bioengineering / Biomedical Engr.**			
Educational Services	$45,800	Wholesale Trade	$66,100
Manufacturing	$46,900	Federal, State & Local Government	$68,800
Professional, Scientific, & Technical Services	$47,500	▪ **Electrical / Electronics Engineering**	
Federal, State & Local Government	$52,300	Professional, Scientific, & Technical Services	$63,200
▪ **Chemical Engineering**			
Manufacturing	$59,800	Manufacturing	$63,500
Professional, Scientific, & Technical Services	$62,200	Federal, State & Local Government	$72,400
		Information	$65,000
Educational Services	$57,800	Educational Services	$62,600
Federal, State & Local Government	$68,200	▪ **Industrial / Manufacturing Engineering**	
Mining, Quarrying, & Oil & Gas Services	$60,400		
▪ **Civil Engineering**		Professional, Scientific, & Technical Services	$56,700
Professional, Scientific, & Technical Services	$57,000	Manufacturing	$55,900
Federal, State & Local Government	$60,600	Federal, State & Local Government	$58,000
Construction	$56,900	Finance and Insurance	$56,900
Educational Services	$56,600	Management of Companies & Enterprises	$58,000
Manufacturing	$54,300		
▪ **Computer Engineering**		Information	$57,500
Professional, Scientific, & Technical Services	$69,800	▪ **Mechanical Engineering**	
		Manufacturing	$61,800
Manufacturing	$74,900	Professional, Scientific, & Technical Services	$64,700
Information	$74,600		
Finance and Insurance	$74,000	Federal, State & Local Government	$72,000
Management of Companies & Enterprises	$72,600	Educational Services	$61,600
		Management of Companies & Enterprises	$66,400
Wholesale Trade	$73,000		

(*continued*)

Table 3.4 (*continued*)

Curriculum / Type of Employer*	2013	Curriculum / Type of Employer*	2013
▪ **Petroleum Engineering**		Wholesale Trade	$75,500
Mining, Quarrying, & Oil & Gas Services	$93,100	Management of Companies & Enterprises	$75,200
Professional, Scientific, & Technical Services	$85,600	▪ **Electrical / Electronics Engineering**	
Manufacturing	$100,500	Professional, Scientific, & Technical Services	$66,400
Management of Companies & Enterprises	$98,100	Manufacturing	$68,300
		Educational Services	$66,300
		Information	$69,700
Master's Degrees		Federal, State & Local Government	$75,300
Computer Science		▪ **Mechanical Engineering**	
Professional, Scientific, & Technical Services	$72,100	Manufacturing	$66,000
Information	$77,300	Professional, Scientific, & Technical Services	$67,600
Manufacturing	$79,000	Educational Services	$67,000
Finance and Insurance	$76,600	Federal, State & Local Government	$75,900

Reprinted from the April 2013 *Salary Survey*, with permission of the National Association of Colleges and Employers, copyright holder.

*Definitions of Employer Types:

Construction: The construction sector comprises establishments primarily engaged in the construction of buildings or engineering products (e.g., highways and utility systems.)

Educational Services: This sector comprises establishments that provide instruction and training in a wide variety of subjects. The instruction and training is provided by establishments such as schools, colleges, universities, and training centers.

Federal, State, and Local Government: Federal, State, and Local Government, excluding state and local schools and hospitals, and the U.S. Postal Service.

Finance and Insurance: This sector includes establishments primarily engaged in financial transactions and/or in facilitating financial transactions.

The Information sector comprises establishments engaged in the following processes: (a) producing and distributing information and cultural products, (b) providing the means to transmit or distribute these products as well as data or communications, and (c) processing data.

Management of Companies and Enterprises: This sector comprises 1) establishments that hold the securities of companies and enterprises for the purpose of owning a controlling interest or influencing management decisions, or 2) establishments that administer, oversee, and manage establishments of the company or enterprise.

Manufacturing: The manufacturing sector comprises establishments engaged in the mechanical, physical, or chemical transformation of materials, substances, or components into new products.

Mining, Quarrying, and Oil and Gas Extraction: This sector comprises establishments that extract naturally occurring mineral solids, such as coal and ores; liquid minerals, such as crude petroleum; and gases, such as natural gas.

Professional, Scientific, and Technical Services: This sector comprises establishments that specialize in performing professional, scientific, and technical activities for others. These activities require a high degree of expertise and training. (Examples - accounting, bookkeeping, payroll services, architectural, engineering, and computer services.)

The Wholesale Trade sector comprises establishments engaged in wholesaling merchandise, generally without transformation, and rendering services incidental to the sale of merchandise. The merchandise described in this sector includes the outputs of agriculture, mining, manufacturing, and certain information industries, such as publishing.

Source: U.S. Census Bureau, North American Industry Classification System (NAICS), 2007. www.census.gov/eos/www/naics/

One question often raised by engineering students concerns the long-term earning potential for engineers. In other words, does the high starting salary hold up over the length of one's engineering career? Figure 3.7 presents 2012 median salaries for all engineers based on the number of years since their bachelor's degree. It is apparent that an engineer's relative earnings hold up well for many years following graduation.

One significant difference in salaries is shown in Figures 3.8 and 3.9. These data relate the 2012 salary difference between engineers who are placed into supervisory positions and those who remain in non-supervisory roles. Over the course of their career, supervisors tend to earn significantly higher salaries than those in non-supervisory positions.

Figure 3.10 shows how an engineer's income is influenced by the level of education achieved. As might be expected, those with master's and doctoral degrees have higher compensation levels. Length of experience in the engineering field is also an income determinant. Figure 3.10 demonstrates a consistent, regular growth in an engineer's income level as experience is gained. Generally, the average increase is at about 4.6% per year of experience.

Table 3.5 demonstrates 2012 median salaries by region of employment. This compares graduates of 1990–1993 with those of 2000–2001. For 10-year-out graduates, salaries are highest in the Mountain Region. The lowest reported salaries are for engineers in the

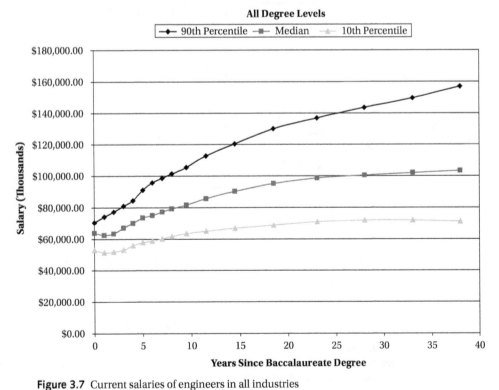

Figure 3.7 Current salaries of engineers in all industries

Source: "Engineers Salaries: Special Industry Report, 2012." Engineering Workforce Commission of the American Association of Engineering Societies, Inc.

Table 3.5 Median Income by Region

Region	Graduates of 2000–2001	Graduates of 1990–1993
New England (CT, ME, MA, NH, RI, VT)	85,690	98,422
Middle Atlantic (NJ, NY, PA)	82,474	92,379
East North Central (IL, IN, MI, OH, WI)	85,282	100,374
West North Central (IA, KS, MN, MO, NE, ND, SD)	81,679	105,461
South Atlantic (DE, DC, FL, GA, MD, NC, SC, VA, WV)	84,736	98,979
East South Central (AL, KY, MS, TN)	74,963	91,901
West South Central (AR, LA, OK, TX)	80,972	82,501
Mountain (AZ, CO, ID, MT, NM, NV, UT, WY)	88,829	105,981
Pacific (AK, CA, HI, OR, WA)	85,177	99,172

Source: 2012 Engineers' Salaries: Special Industry Report
Engineering Workforce Commission of the American Association of Engineering Societies, Inc.

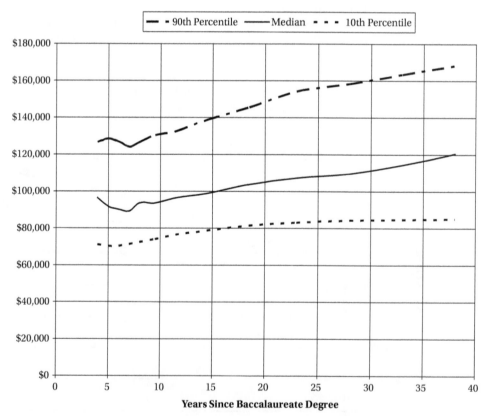

Figure 3.8 Salary curves for engineering supervisors, all degree levels
Source: "Engineers' Salaries: Special Industry Report, 2012." Engineering Workforce Commission of the American Association of Engineering Societies, Inc.

Figure 3.9 Salary curves for engineering non-supervisors, all degree levels

Source: "Engineers' Salaries: Special Industry Report, 2012." Engineering Workforce Commission of the American Association of Engineering Societies, Inc.

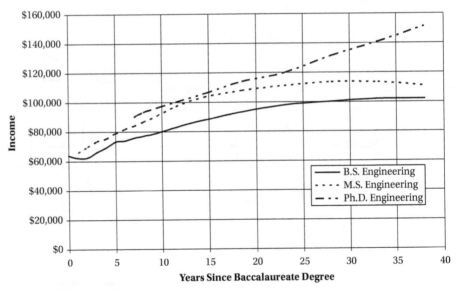

Figure 3.10 Current median income of all engineers in all industries by highest degree earned and length of experience

Source: "Engineers' Salaries: Special Industry Report, 2012." Engineering Workforce Commission of the American Association of Engineering Societies, Inc.

East South Central Region. (Geographic strongpoints have shifted dramatically and repeatedly in recent years.)

For 20-year-out graduates, salaries are highest in the Mountain and West North Central Regions. Lowest salaries are in the West South Central Region.

Figure 3.11 shows the current median income of all engineers in all industries by the size of the company and length of experience. Large companies are those with over 5,000 total employees. Medium companies have between 500 and 5,000 total employees. Small companies are those with 500 or fewer total employees.

For the first 20 years, all groups are quite similar by income. However, at the 20-year point, the incomes of those with medium companies increase significantly higher than those of the large and small companies.

Figure 3.12 provides the 2011 median annual income by major branch of engineering for all working engineers. Petroleum engineers show the highest income levels, with agricultural engineers at the lowest levels.

It is also interesting to examine median annual income by industry or service. Table 3.6 compares the 2012 median salaries of graduates from 1990–1993 with those of 2000–2001. The data show those involved in certain manufacturing areas are at the top of the salary list. The lowest from these classes are involved in public administration.

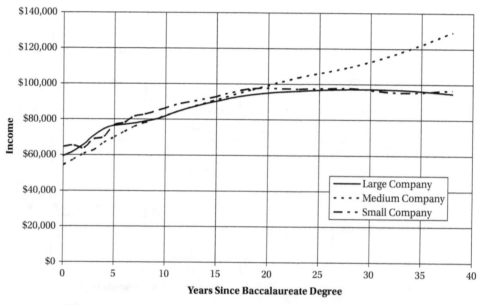

Figure 3.11 Current median income of all engineers in all industries by size of company and length of experience

Source: "Engineers' Salaries: Special Industry Report, 2012." Engineering Workforce Commission of the American Association of Engineering Societies, Inc.

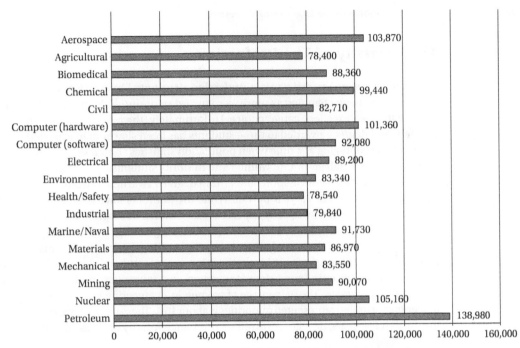

Figure 3.12 Income by engineering discipline, 2011 (mean annual wage)

Source: Bureau of Labor Statistics, Division of Occupational Employment Statistics, 2011

Table 3.6 Median Income by Industry Sector

Industry Sector	Graduates of 2000–2001	Graduates of 1990–1993
All Manufacturing Industries	87,929	100,659
Metal Manufacturing Industries	N/A	88,751
Metal Products Manufacturing	73,931	81,631
Machinery Manufacturing	74,932	89,979
Transportation Equipment Manufacturing	94,540	100,130
Motor Vehicle Manufacturing	96,014	100,754
All Non-Manufacturing Industries	78,433	93,281
Transportation and Warehousing	73,231	84,196
Architectural, Engineering, and Related Services	82,246	106,204
Construction	82,174	100,082
Mining	86,471	107,730
Electric Power Generation	75,655	103,106
Scientific Research and Development Services	84,951	104,269
Professional, Scientific, and Technical Services	84,083	103,682
Public Administration	70,583	88,875
Utilities	79,161	103,900

Source: 2012 Engineers' Salaries: Special Industry Report

Engineering Workforce Commission of the American Association of Engineering Societies, Inc.

3.7   The Diversity of The Profession

For many years, engineering was a profession dominated by white males. In recent years, this has been changing as more women and minority students have found engineering to be an excellent career choice. Unfortunately, the rate of change in both enrollment and degrees awarded to these students has been slower than expected. This is especially discouraging at a time when population statistics show a significant increase in the number of college-bound women and minority students.

Table 3.7 provides information concerning degrees awarded to women and minority groups over a recent 10-year period. While the numbers have been increasing, the growth rate has been slower than expected. Since recent demographic data show that the number of minority students attending college is growing, we would expect to see a larger percentage of these students earning engineering degrees. So far, this is not the case—especially at the graduate level, where increases have been steady, but modest.

Table 3.8 shows enrollment data for women and minorities in all undergraduate engineering programs. The percentage of women in engineering is over 18%, while the percentage of under-represented minorities is about 15%. These figures have dropped during the past few years, despite increased recruitment efforts by colleges and universities.

3.8   Distribution of Engineers by Field of Study

According to the Bureau of Labor Statistics (Fig. 3.13), over 290,000 of all practicing engineers are employed in the electrical engineering discipline. This is followed by about

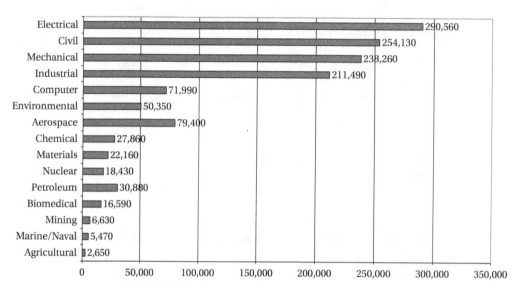

Figure 3.13 Practicing engineers in the U.S., 2012
Source: Bureau of Labor Statistics, Occupational Employment Statistics, March, 2012

Table 3.7 Degrees in Engineering, by Level, and Type of Student, 2002–2011

Level and Type	2002	2003	2004	2005	2006	2007	2008	2009	2010	2011
Bachelor's Degrees	68,648	75,031	76,003	76,003	76,103	75,823	77,107	75,320	79,528	84,599
Women	14,102	15,114	15,282	14,868	14,654	14,101	13,865	13,432	14,478	15,437
African Americans	3,358	3,429	3,699	3,756	3,673	3,735	3,470	3,392	3,635	3,457
Hispanic Americans	4,298	4,652	4,813	4,890	4,957	5,133	5,486	5,488	6,053	6,574
Native Americans	315	388	362	378	456	426	427	328	378	363
Asian Americans	8,669	9,705	9,941	10,033	9,719	9,466	9,143	8,743	9,174	9,680
Foreign Nationals	4,859	5,560	5,768	5,644	5,354	5,152	4,787	4,519	4,948	5,532
Master's Degrees	31,983	36,611	40,953	41,087	38,451	37,805	40,122	41,967	43,257	45,589
Women	7,067	8,024	8,842	9,212	8,731	8,393	9,237	9,596	9,798	10,310
African Americans	867	916	947	1,072	1,009	1,078	1,116	1,137	1,219	1,246
Hispanic Americans	817	927	1,143	1,194	1,185	1,292	1,318	1,393	1,543	1,617
Native Americans	72	89	91	123	128	81	78	94	98	105
Asian Americans	2,829	3,172	3,934	3,994	3,990	3,995	3,863	3,629	4,097	4,316
Foreign Nationals	13,615	16,831	18,447	17,536	15,441	14,604	16,408	18,426	18,360	18,868
Doctoral Degrees	5,863	6,027	6,504	7,276	8,116	8,614	9,449	8,907	9,243	10,086
Women	1,024	1,040	1,136	1,322	1,592	1,689	1,976	1,896	2,073	2,185
African Americans	94	97	102	111	121	122	127	139	165	169
Hispanic Americans	112	107	97	107	99	120	152	163	224	206
Native Americans	8	13	7	15	11	15	15	16	14	12
Asian Americans	401	337	397	391	496	432	508	582	678	839
Foreign Nationals	3,217	3,441	3,766	4,405	5,048	5,180	5,564	4,869	4,871	5,334

Source: 2002–2011: Engineering Workforce Commission of the American Association of Engineering Societies

Table 3.8 Women and Minorities in the Engineering Pipeline, 2011

	First-Year Full-Time B.S. Students	All Full-Time B.S. Students
Women	18.45%	18.61%
African Americans	6.30	5.13
Hispanic Americans	10.49	9.85
Native Americans	0.54	0.49
Asian Americans	9.35	10.78

Data represent percentage of all Full-Time Undergraduates.
Source: Engineering Workforce Commission of the American Association of Engineering Societies

254,000 civil engineers and 238,000 mechanical engineers. It is interesting to note that "other engineers" (which includes the smaller, specialized fields of engineering) include about 250,000 engineers. Comparing this information to previous data on enrollment and degrees, one would expect these numbers to remain somewhat steady.

3.9 Engineering Employment by Type of Employer

According to the Bureau of Labor Statistics, in 2008 the largest sector of engineering employment, almost 40%, was in manufacturing industries. These include electrical and electronic equipment, industrial machinery, transportation equipment, and instruments.

In 2008, over 480,000 engineering jobs were in service industries which includes engineering and architectural services, research and testing services, and business services in which firms designed construction projects or did other engineering work on a contractual basis. Engineers were also working in communications, utilities, and computer industries.

Government agencies at the federal, state, and local levels employed approximately 192,000 engineers in 2008. The largest number were employed by the federal government in the Departments of Defense, Transportation, Agriculture, Interior, Energy, and the National Aeronautics and Space Administration (NASA). The majority of state and local workers are usually involved in local transportation, highway, and public works projects.

In 2000, almost 48,000 engineers were self-employed, typically as consultants.

3.10 Percent of Students Unemployed or in Graduate School

How does engineering unemployment compare to the number of students who go on to graduate school? As indicated in Table 3.9 and referred to earlier in Figure 3.1, the total number of students pursuing graduate engineering degrees increased significantly between the late 1970s and 1992. Since then, there had been a steady decline until 2000. During the period 2001–2011 enrollment in graduate engineering programs has soared to 168,000. The recent trend is similar to the early period covered by Table 3.9 when jobs for bachelor-level graduates were not as plentiful, and many are opting for graduate school, which may explain the rapid rise in graduate school enrollments.

Table 3.9 also shows the lower unemployment rate for graduating engineers between 1990 and 2008. The figures for 1992 and 1994 and 2002–2011 reflect a tighter job market than that of recent years. It should also be noted that, in general, the unemployment rate for new engineering graduates is far lower than for any other major.

As the engineering job market begins to improve, one might expect that the unemployment rate for engineers may decrease, and the number going to graduate school may also start to decline.

3.11 A Word from Employers

As discussed earlier, the current job market for graduating engineers is challenging. However, there are important issues that employers want students to consider. Even though this study is over 14 years old, the message is still important for today's engineering college student. In a 1999 survey of 450 employers done by the Collegiate Employment Research Institute at Michigan State University, recruiters were asked to provide commentary on any special concerns they had about new engineering graduates.

One interesting conclusion to the study found that "Employers want the total package when they hire their next engineering graduates. Not satisfied with academically well-prepared graduates, employers want individuals who possess and can demonstrate excellent communication and interpersonal skills, teamwork, leadership, and computer/technical proficiency. A willingness to learn quickly and continuously, to problem solve effectively, and to use their common sense is also desired. New employees must be hard-working, take initiative, and be able to handle multiple tasks." In addition, employers are looking for students with new emerging skills and aptitudes such as the ability to understand e-commerce, computer capabilities that include programming skills, and the ability to adapt to constant change. In the new global economy, increasing competition requires a strategy to respond quickly.

Therefore, it is important to realize that just earning an engineering degree is no longer sufficient to prepare students for the fast-paced, technologically changing global economy. Students must devote time in college developing those critical competencies that employers are seeking in order to be well prepared for the challenges that await them.

Table 3.9 Selected Disciplines' Engineering Graduate School Full- and Part-time Enrollment Trends and Percent of Unemployed Engineers

	1990	1992	1994	1998	2000	2002	2004	2005	2007	2008	2011
Aerospace	3,563	3,935	3,550	2,780	3,042	3,171	3,750	3,758	3,990	4,291	5,250
Chemical	6,657	6,927	7,292	6,683	6,714	6,963	7,282	6,878	6,949	7,268	8,073
Civil	77,625	13,900	14,164	11,009	11,151	12,546	12,942	12,561	12,787	13,430	16,545
Electrical	32,095	32,635	29,524	25,302	27,764	33,851	34,127	32,691	35,306	35,778	35,750
Industrial	9,713	11,552	11,401	9,428	5,234	6,330	5,789	5,340	5,677	6,302	6,913
Mechanical	17,115	18,566	17,310	14,032	15,201	17,085	17,945	17,554	18,349	18,880	21,860
Total	80,750	87,510	83,230	69,220	69,106	79,946	81,835	78,782	83,058	85,949	94,391
% Unemployed Engineers	2.9%	5.2%	4.3%	1.7%	1.5%	3.0%	2.3%	2.1%	3.3%	3.1%	3.1%

Sources: "Engineering Enrollments and Degrees," Engineering Manpower Commission of the
American Association of Engineering Societies, Inc. 1990, 1992, 1994, and Engineering Workforce
Commission of the American Association of Engineering Societies, Inc., 1998, 2000, 2002, 2004, 2005, 2007, 2008, 2011.

EXERCISES AND ACTIVITIES

3.1   If an administrator used only the data from 1975 to 1980 to predict future engineering enrollment, how many first-year, full-time students would have been expected in 2010? What would the percentage error have been?

3.2   What is the percentage increase in projected high school graduates in 2021 compared to those in 2010?

3.3   Using the data of Figure 3.5, estimate the number of employed engineers in 2015.

3.4   Using the starting salary data from Table 3.3, and assuming an annual increase at the current cost of living (currently about 2%), calculate what you could expect to receive as a starting salary in your chosen field of study if you graduate three years from now.

3.5   Using the information from Figure 3.10, estimate the engineering salary after 30 years assuming a) a bachelor's degree; b) a master's degree; and c) a doctorate degree.

3.6   Using the information from Figures 3.8 and 3.9, estimate the median salary after 30 years assuming the engineer is a) a supervisor and b) a non-supervisor.

3.7   If you were a supervisor in the upper decile rather than a non-supervisor in the upper decile, what would be your percentage increase in salary 20 years after graduation?

3.8   Calculate the percentage increase of the highest-paid bachelor's engineer in 2013 compared to the lowest-paid bachelor's engineer in 2013.

3.9   Suggest at least one reason for the relatively low income in the East South Central Region of Table 3.5.

3.10   Suggest at least one reason for the relatively low income of government or public administration workers compared to those in the private sector.

3.11   Review the data in Tables 3.7 and 3.8. As the enrollment of women and minority students are rising on college campuses, the relative growth for these groups studying engineering has not been increasing as quickly. Discuss some reasons, and provide examples, why you feel that more women and minorities are not pursuing engineering degrees. Explore some of the student support programs on your campus. What are some of the things your school is specifically doing to encourage and support these populations to pursue engineering degrees?

3.12   Estimate the percentage of engineers that are chemical engineers using the data of Figure 3.13.

3.13   Based on the current economic situation, do you expect the employment demand for graduating engineers to increase or decrease? Explain the basis for your answer.

3.14   If we experience a significant economic recovery, what do you think will happen to enrollment in graduate engineering programs?

3.15   Review the information presented in Figures 3.2 and 3.3 and Table 3.7. Make a prediction about the future enrollment of women and minorities in engineering. Explain your answer.

3.16   What factors do you think influence how employers determine hiring needs in any given year?

3.17   Attend a campus career fair. Talk with at least 3 recruiters from different firms. Make a list and discuss those factors which employers are stressing in their hiring decisions.

3.18   Based on the current economic situation, do you expect the salaries of engineers to continue to increase at the recent pace? Why or why not? Using other tables in this chapter, explain the basis for your answer.

3.19   Review the enrollment data in this chapter. What do you think high schools, colleges, and the engineering profession can do to increase the number of students choosing to pursue an engineering career?

3.20   What factors do you think contribute to the fact that entry-level engineering salaries are usually higher than any other major on campus?

CHAPTER 4

Succeeding in the Classroom

4.1 Introduction

The methods for excelling in the classroom outlined in this chapter have proven effective for numerous students. We suggest that you consider all of these ideas and adapt the ones that fit best with your own personality. If you are unsure whether they will work, try some out for a trial period and see. If those don't work, move on to others until you find something that fits with your own style.

The three components to succeeding in your academics are your ability, your attitude, and your effort. There is no book that can address the first item. While there is unquestionably a level of ability that is needed to succeed in engineering, there are numerous students who had the ability but failed at engineering because of poor attitudes or poor or inefficient effort. There are also many students who lacked natural ability but because of their positive attitude and extraordinary effort are now excellent engineers. This chapter focuses on techniques to improve these two components for success that are totally under your control.

4.2 Attitude

"Ability is what you're capable of doing. Motivation determines what you do. Attitude determines how well you do it."

—Lou Holtz, South Carolina football coach

Your attitude is the first thing that you have control over when beginning any challenge, including studying engineering. You can expect success or expect failure. You can look for the positives or dwell on the negatives.

Approaching your classes, professors, and teaching assistants with a positive attitude is the first key to succeeding in your engineering studies. Each class should be looked at as an opportunity to succeed. If there are difficulties along the way, learn to deal with those and move ahead. Many students will decide that a certain class is too difficult or that the professor is a poor teacher and will expect to do poorly. Many fulfill their own prophecy and actually fail. Other students look at the same situation and overcome the hurdles and excel.

4.3   Goals

In setting goals, there are two essential elements—"height" and time. You might ask, "How high do I set my goals?" This will vary from individual to individual. Goals should be set high, but attainable. A common management practice is for a team to decide on a reasonable goal to accomplish. Then the bar is raised and a stretch goal is defined that encourages the team to produce more. This stretch goal is designed to push the group farther than they would have gone with only the original goal. Sometimes this stretch goal looks reasonable and sometimes it looks totally unreachable. What often happens, though, is that the stretch goal is met, realistic or not. As an individual, setting stretch goals helps us grow. When setting goals for yourself, look at what you think you can do and set base goals. Then establish stretch goals a little higher than the base goals. You may just find yourself meeting those higher goals.

The second key element in goal setting is time. When will the goal be accomplished? It is important to have long-term goals and short-term goals. Long-term goals help guide where you are headed. For example, what do you want to do with your engineering degree? An appropriate long-term goal as a freshman could involve determining the kind of entry-level job you desire. This will help answer the question "Why am I doing this?" when your classes get tough.

Short-term or intermediate goals should be stepping-stones to the achievement of your long-term goals. An advantage of an academic setting is that there are natural breaks, semesters, or quarters, for the establishment of goals within specific time frames. Goals can be set for the academic year, and for each semester or quarter. These provide intermediate goals.

Short-term goals need to be set and rewarded. Rewarding short-term goals is essential to keeping motivation high in the quest for long-term goals. Many students lose sight of their long-term goals because they didn't set short-term, "rewardable" goals. Short-term goals may be an "A" on an upcoming quiz or test, or it may be completing a homework set before the weekend.

Clearly define your goals by writing them down. It is much more powerful to write down your goals than to simply think about them. Some people post them where they can see them daily. Others put them in a place where they can retrieve and review them regularly. Either method is effective. For short-term goals, establish weekly goals. Select a day, probably Sunday or Monday, and establish the goals for the week. This is a great time to establish the rewards for your weekly goals. Plan something fun to do on the weekend if you meet those goals.

At the end of each semester, examine your progress toward your long-term goals. This is an advantage to being in school. Every semester, there is an opportunity to assess your progress and start fresh in new classes. Take advantage of this. Examine your long-term goals and make any changes if necessary.

Determining Goals

It can be intimidating or difficult for students early in their careers to decide on appropriate goals. Fortunately, there are numerous resources on campus to help. Academic advisors can provide criteria for grade-point goals. Things to consider include:

1. What is the minimum GPA to stay enrolled in school?
2. What grades do you need to continue in engineering or to enter engineering?
3. What grades are needed to be eligible for co-op or intern positions?
4. What GPA is required to make the dean's list or honor list?
5. What GPA is needed to be eligible for the engineering honorary organizations?
6. What grades are needed for admittance to graduate school?
7. What grades are needed for scholarships?

There are also people on campus who work with the placement of graduates. They are excellent resources to get information on what employers are looking for. What credentials do you need to be able to get the kind of job you desire? If you know this as a freshman, you can work toward it. You can also go directly to the employers and ask these same questions of them. If companies come to your campus, talk to them about what credentials are needed to get a job. Job fairs are great times to do this. Approaching company representatives about establishing goals is also a great way to show initiative and distinguish yourself from your classmates.

Goals are important for outside the classroom too. It is important to have goals in all areas of your life. Later in this chapter we will discuss such goals.

4.4   Keys to Effectiveness

Once goals have been set regarding grades, the next step is to achieve those grades. Effort and effectiveness may be the most important components of your success as a student. There are numerous cases of well-equipped students who end up failing. There are also students who come to the university poorly prepared or start off poorly but eventually succeed. The difference is in their effort and their effectiveness. The following suggestions contain strategies to improve personal effectiveness in your studies.

> A general rule is that it takes a minimum of two hours of study outside of class for each hour of class lecture.

Take Time to Study

One of the strongest correlations with a student's performance is the time he or she spends studying.

According to the reminder above, if you are taking 15 hours, that means that you need to spend 30 effective hours a week studying outside of class. In a typical high school, the time requirement is much less than this. Developing the study habits and discipline to spend the needed time studying is a prime factor in separating successful and unsuccessful students.

Go to Class

There is also a high correlation with class attendance and performance. This may seem obvious, but you will run into numerous students who will swear that it doesn't matter if you go to class. Many instructors in college do not take attendance, so no one but you knows if you were there, which is different from high school. The classroom is where important information will be presented or discussed. If you skip class, you may miss this information. Also, by skipping class, you miss an opportunity to be exposed to the material you will need to know, and you will have to make up that time on your own. Missing class only delays the expenditure of time you will need to master the material. Most often, it may take much more than an hour of independent study to make up an hour's worth of missed class time. It is possible for you to never miss a class in your undergraduate years, except for illness or injury, serious family problems, or trips required by groups to which you belong. Never missing a class could be one of your goals.

Also, you may want to schedule classes at 8 a.m. and 4 p.m. This will get you up and going and keep you there! You will seldom, if ever, study at 8 a.m. and at 4 p.m. Do your studying in the middle of the day. This will help you to use your time effectively.

> "After class, reread the assigned sections in your textbook. Concentrate on the sections highlighted by the lecture. By doing so, you will not waste time trying to understand parts of the book that are not critical to the course."

Make Class Effective

The first component of making class effective is sitting where you can get involved. If the classroom is large, can you see and hear in the back? If not, move up front. Do you need to feel involved in the class? Sit in the front row. Do you fall asleep in class? Identify why. Ask yourself, "Am I sleeping enough at night?" If the answer is no, get more sleep.

The second step for making class time effective is to prepare for class. Learning is a process of reinforcing ideas and concepts. As such, use the class time to reinforce the course material by reviewing material for the lecture beforehand. Most classes will have a textbook and assigned reading for each lecture. Take time before the lecture to skim over the relevant material. Skimming means reading it through but not taking time to understand it in depth. This will make the class time more interesting and more understandable, and it will make note taking easier.

By following the above advice, you will have had three exposures to the material, which helps you remember and understand the material better.

Keep Up with the Class

Class is most effective if you keep up. An excellent short-term goal is to master the lecture material of each course before the next lecture. A course is structured in such a way that you master one concept and then move on to the next during the next lecture. Often in science and engineering classes the concepts build on each other. The problem with falling behind is that you will be trying to master concepts that depend on previous lectures, and you are going to hear lectures that you will not understand.

There is a very practical reason that a three-credit class meets three times per week. During a 15-week semester, there are only 45 hours of class meeting time. That amount of class time could be held in a week of consecutive nine-hour days. Arranged that way, a whole semester could be taken in a month. Then it would only take a year to get a bachelor's degree. So why don't we do it that way?

The reason is that the current educational model is designed to introduce a concept and give you, the learner, time to digest it. It is an educationally sound model. In engineering study, digesting means reviewing notes, reading the text, and working several problems related to the topic. If you are still confused about an aspect of the previous lecture, the next class provides a great opportunity to bring it up with the professor. Get your questions answered before going on to the next topic.

A note of caution should be given here. Friends of yours may be studying subjects where they can study from test to test rather than from class to class. They may insist that you don't need to keep up so diligently because they don't. Some majors cover material that is more conceptually based. That is, once you understand the concept, you have it. Engineering, math, and science courses are not this way. To be mastered they require extensive study and preparation over an extended period of time. In some courses there are only a few basic concepts, yet the entire course is dedicated to the application of these few concepts. An example is Statics, a sophomore-level course, in which there are only two fundamental equations germane to the course: the sum of forces equals zero, and the sum of moments equals zero. The entire course involves applying these two equations to various situations, and using the information in engineering applications.

Take Effective Notes

A main component in keeping up with your classes is taking effective notes for each lecture. Effective notes capture the key points of the lecture in a way that allows you to understand them when you are reviewing for the final exam three months later. Suggested note-taking strategies follow:

1. Skim the assigned reading prior to class to help identify key points.
2. Take enough notes to capture key points but don't write so much that you fail to adequately listen to the presentation in class.
3. Review your notes after class to annotate them, filling in gaps so they will still make sense later in the semester.
4. Review your notes with other students to ensure that you captured the key points.
5. Review your notes early in the semester with your professor to be certain you are capturing the key points.

In class, your job is to record enough information to allow you to annotate your notes later. You don't have to write everything down. Getting together with classmates after class is a terrific way to annotate your notes. That way you have different perspectives on what the main points were. If you really want to make sure you captured the key points, go and ask your professor. Doing so early in the semester will not only set you up for successful note taking throughout the semester, it will also allow you to get to know your professor.

Work Lots of Problems

Because the applications of the concepts are the core of most of your classes, the more problems you can work, the better prepared you will be. In math, science, or engineering courses, doing the assigned homework problems is a minimum. Search for additional problems to work. These may be from the text or from old exams. Your professor should be able to steer you to appropriate problems to supplement the homework.

Caution Using Solution Manuals and Files

For some classes, there are homework files or solution manuals available. While these tools may help get the homework done quicker, they very often adversely affect exam performance. The problem is that most professors don't allow the solutions to be used during the test. Using them becomes a crutch and can impair your ability to really learn the material and excel on exams. If they are used at all, use them only after exhausting all other possibilities of working out the answer yourself. If it takes a lot of work to figure out a homework problem, that concept will be learned well and you will have a higher probability of demonstrating your knowledge on the next test. Taking the easy way out on homework has a very consistent way of showing up on test results. Minimal learning results from copying down a solution.

Group Studying

A better model for studying than using solutions is to study in a group. Numerous studies have shown that more learning takes place in groups than when students study by themselves. Retention is higher if a subject is discussed, rather than just listened to or read. If you find yourself doing most of the explaining of the ideas to your study partners, take heart! The most effective way to learn a subject is to teach it. Anyone who has taught can confirm this anecdotally; they really learned the subject the first time they taught it.

> *"The most important single ingredient to the formula of success is knowing how to get along with people."*
>
> —Theodore Roosevelt

If you aren't totally convinced of the academic benefits of group studying, consider it a part of your engineering education. Engineers today work in groups more than ever before. Being able to work with others effectively is essential to being an effective engineer. If you spend your college years studying in groups, working in groups will be second nature when you enter the workforce.

Studying with others will also make it more bearable, and even fun, to study. There are numerous stories about students preparing for exams and spending the whole day on a subject. It is difficult to stick to one subject for an entire day by yourself. With study partners you will find yourself sticking to it and maybe even enjoying it.

Studying with others makes it easier to maintain your study commitments. Something we all struggle with is discipline. It is much easier to keep a commitment to study a certain subject if there are others who are depending on you. They will notice if you are not there and studying.

> *"I will pay more for the ability to deal with people than any other ability under the sun."*
>
> —John D. Rockefeller

Group studying is more efficient if properly used. Chances are that not everyone in the group will be stuck on the same problems. So right away, you will have someone who can explain the solution. The problem areas common to everyone before the group convenes can be tackled with the collective knowledge and perspectives of the group. Solutions are found more quickly if they can be discussed. More efficient study time will make more time for other areas of your life!

Choosing a group or partner may be hard at first. Try different study partners and different size groups. Ray Landis, Dean of engineering at California State University, Los Angeles and a leading expert on student success, suggests studying in pairs. "That way each gets to be the teacher about half the time." Larger groups also can be effective. The trick is to have a group that can work efficiently. Too large a group will degenerate into a social gathering and not be productive in studying. Whichever group size you choose, here are some basic tips for group studying.

1. Prepare individually before getting together.
2. Set expectations for how much preparation should be done before getting together.
3. Set expectations on what will be done during each group meeting.
4. Find a good place to convene the group that will allow for good discussion without too many distractions.
5. Hold each other accountable. If a group member is not carrying his or her weight, discuss it with that person and try to get them to comply with the group's rules.
6. If a member continues to fail to carry their own weight or comply with the group's rules, remove that person from the group.

Select a Good Study Spot

In a subsequent section, differences in learning styles will be discussed. Depending on your own learning style and personality, you may need a certain kind of environment to study efficiently. Some resources attempt to describe the "perfect study environment." In reality, there is no one perfect study environment or method of studying. What is crucial is that you find what you need and find a place that meets your needs. For some, it involves total quiet, sitting at a desk. Others may prefer some noise, such as background music, and prefer to spread out on the floor. Whatever you decide is the best environment, pick one where you can be effective.

Cynthia Tobias describes the conflict she has with her husband's view of a proper study environment in her book *The Way They Learn:*

I have always favored working on the floor, both as a student and as an adult. Even if I'm dressed in a business suit, I close my office door and spread out on the floor before commencing my work. At home, my husband will often find me hunched over books and papers on the floor, lost in thought. He is concerned.

"The light is terrible in here!" he exclaims. "And you're going to ruin your back sitting on the floor like that. Here, here! We have a perfectly clean and wonderful rolltop desk." He sweeps up my papers and neatly places them on the desk, helps me into the chair, turns on the high-intensity lamp, and pats my shoulder.

"Now, isn't that better?" he asks.

I nod and wait until he is down the hall and out of sight. Then, I gather all my papers and go back down on the floor. . . . It does not occur to him that anyone in their right mind could actually work better on the floor than at a desk, or concentrate better in 10-minute spurts with music or noise in the background than in a silent 60-minute block of time.

Different people have different ways in which they can be most efficient. You need to discover yours so that you can be efficient. If you are unsure of what works best for you, test some options. Try different study environments and keep track of how much you get done. Stick with the one that works best for you.

If you and your roommate have different styles and needs, then you will need to negotiate to decide how the room will be used. Both of you will have to be considerate and compromise, because it is a shared space.

One item that is a distraction for almost everyone is television. Because it is visual, it is almost impossible to concentrate on homework with a TV on, although many students swear they are still "productive." A few may be, but most are not. Watching TV is one of the biggest time-wasters for students. It is hard enough for first-year students to adjust to college life and the rigors of pre-engineering classes. We strongly recommend ***not*** having a television set in your residence hall room or apartment for at least your first year. It is not even good for study breaks, which should be shorter than 30- or 60-minute programs. This may be one temporary sacrifice you have to make to be successful in your studies.

4.5 Test-Taking

Most courses use written, timed exams as the main method of evaluating your performance. To excel in a course, you need to know the material well and be able to apply it in a test situation. The early part of this chapter provided tips on how to improve your understanding of the course material. The second part is test-taking skills. Ask any student who has been away from school for a few years. Taking tests is a skill and needs to be practiced. Just like a basketball player who will spend hours practicing lay-ups, it is important to practice taking tests. A great way to do this is to obtain past exams. A couple of days before the test, block out an hour and sit down and "take" one of the old tests. Use only the materials allowed during the actual test. For instance, if it is a closed-book test, don't use your book. In this way, you are doing two things. The first is assessing your preparation. A good performance on the practice test indicates that you are on the right track. The second thing you are doing is practicing the mechanics of taking the test. Again, with the basketball analogy, players will scrimmage in practice to simulate game conditions. You should simulate test conditions. What do you do when you reach a problem you can't do? Most people will start to panic, at least a little bit. It is critical that you stay calm and reason your way beyond whatever is blocking you from doing the problem. This may mean skipping the problem and coming back to it, or looking at it in another way. In any case, these are the types of things you have to do on tests, and by practicing them under test conditions, you will do better on the actual tests.

Before the first test in the course, visit your professor and ask him or her how to assemble a simulated test. It may be that old exams are the way to go, or possibly certain types of problems will be suggested. Professors don't like the question "What is going to be on the test?" but are much more receptive to requests for guidance in your preparation. The professor's suggestions will also help you focus your studying, to make it more efficient.

Taking the Test

There are some general guidelines for taking any test. The first one is to come prepared to take the test. Do you have extra pencils just in case? Can you use your calculator efficiently? Breaking your only pencil or having your calculator fail during the test can be very stressful and prevent you from doing your best, even if you are given a replacement. Proper planning for the test is the first step toward succeeding.

The second step is to skim over the entire test before beginning to work the problems. This is partially to make sure that you have the complete test. It also gives you an overview of what the test is going to be like so you can plan to attack it efficiently. Take note of the weighting of the points for each problem and the relative difficulty.

The next step is to look for an easy problem or one that you know you can do. Do this problem first. It will get you into the flow of doing problems and give you confidence for the rest of the test. Also, doing the problems you know you can do will ensure that you have made time for the problems for which you should get full credit.

Always keep track of the time during a test. Tests are timed, so you need to use the time efficiently. Many students will say they "just ran out of time." While this may prevent you from finishing the test, it should not prevent you from showing what you know. Look at the number of points on the test and the time you have to take the test. This will tell you how long to spend on each problem. If there are 100 points possible on the test and 50 minutes to complete it, each point is allotted 30 seconds. So a 10-point problem should only take 5 minutes. A common mistake is wasting too much time struggling through a difficult problem and not even getting to others on the test. Pace yourself so you can get to every problem even if you don't complete them. Most instructors will give partial credit. If they do, write down how you would have finished the problem and go on to the next when the allotted time for that one is up. If there is time at the end, you can go back to the unfinished problem. But if not, the instructor can see that you understood the material and can award partial credit.

Remember that the goal of a test is to get as many points as possible. A professor can't give you any credit if you don't write anything down. Make sure you write down what you know and don't leave questions blank. Another common mistake occurs when students realize they have made a mistake and erase much of their work. Often, the mistake they made was minor and partial credit could have been given, but none is awarded because everything was erased.

Concentrate on the problems that will produce the most points. These are either problems you can do relatively quickly and get full credit for, or ones worth a large portion of the total points. For instance, if there are four problems, three worth 20 points and one worth 40 points, concentrate on the 40-point problem.

Leave time at the end, if possible, to review your answers. If you find a mistake and there is not enough time to fix it, write down that you found it and would have fixed it if you had time. If you are taking a multiple-choice test, be careful about changing your answers. Many studies find that students will change right answers as often as they change wrong answers. If you make a correction on a multiple-choice test, make certain you know that it was wrong. Otherwise, your initial answer may have been more likely to have been correct.

Think. This may sound basic, but it is important. What is the question asking? What does the professor want to see you do? When you get an answer, ask yourself if it makes sense. If you calculate an unrealistic answer, comment on it. Often a simple math error produces a ridiculous answer. Show that you know it is ridiculous but that you just ran out of time to find the error.

After the Test

After you get your test back, look it over. It is a great idea to correct the problems you missed. It is much easier to fill in the holes in what you missed as you go along in the semester than to wait until you are studying for the final exam. Most final exams cover the material from all the semester tests. If you can't correct the problems yourself, go and see the professor and ask for help. Remember, you *will* see the material again!

4.6 Making the Most of Your Professors

One of the most under-utilized resources at a university is the faculty. Many students do not take full advantage of the faculty that they are paying to educate them. The professor is the one who decides on course material, what is important and what is not, and how you will be evaluated. Yet few students take the time to get to know their professors. Here are some reasons to get to know your professors:

1. They are professionals and have experience in the fields you are studying and, therefore, have perspectives that are valuable to you.
2. Every student will need references (for scholarships, job applications, or graduate school applications) and professors can provide them.
3. They are the ones in charge of the course and can help you focus your studying to be more effective.

4. They are the experts in the field and can answer the hard questions that you don't understand.
5. They assign your grades, and at some point in your college career, you may be on the borderline between grades. If they know you and know that you are working hard, they may be more likely to give you the higher grade.
6. They are likely to know employers and can provide job leads for full-time positions or internships.
7. They may be aware of scholarship opportunities or other sources of money for which you could apply.
8. They can provide opportunities for undergraduates to work in labs, which is great experience. You may even get paid!

It is interesting to examine the history of the university. In centuries past, a student did not go to a university to get "trained" to do a job. They went to be mentored by great scholars (the faculty). In modern universities, many students come looking primarily for the training needed to get a good job, and getting mentored by faculty is low among their priorities. Consider this: a school could send you the course materials at home and have you show up for one day to take your exams. Yet students still come to a university to study. Why? A main reason is for the interaction with the people at the university—faculty, staff, and other students.

If there are so many reasons to get to know professors, why don't most students do so? Many professors don't encourage students to come see them, or don't present a welcoming appearance. Professors may remind students of their parents, people who seem to be out of touch with college students. It may also be that many students don't really understand the advantages of getting to know their professors.

Here are some things to keep in mind when getting to know professors. The first is that they have spent their careers studying the material you are covering in your courses and, therefore, they enjoy it. So it would be a very bad idea to start by telling your calculus professor that you find math disgusting. This is insulting and suggests he has wasted his career.

The second is that professors teach by choice, especially in engineering. Industry salaries are higher than academic salaries for engineering Ph.D.s, so professors teach because they want to. They also probably think that they are good at it. If you are having a problem with an instructor's teaching style, approaching him or her with a problem solving strategy such as the following can be very constructive: "I am having trouble understanding the concepts in class. I find that if I can associate the concepts with applications or examples I understand them better. Can you help me with identifying applications?"

As a group, professors also love to talk, especially about themselves and their area of expertise. They have valuable experience that you can benefit from. Ask them why they chose the field they did. Ask them what lessons they learned as students. You will get some valuable insights and also get to know someone who can help you succeed in your class, and possibly in your career. Steven Douglass, in his book *How to Get Better Grades and Have More Fun* suggests asking each professor, "What is the main objective in the course?"

Faculty members are very busy. They have many demands on their time (research, securing grants, writing, committee assignments, etc.) in addition to your class. These other demands are what keep the university running. So respect their time. It is okay to get to know them but don't keep popping in just to shoot the breeze. Also, when you come with questions, show them the work you have been doing on your own; you only need to get past a hurdle. A busy faculty member won't mind helping a hard-working student clear up an idea or concept. However, showing up and asking how to do the homework can get interpreted as "Do my homework for me." This would not make the positive impression you want.

4.7   Learning Styles

Thinking Skills

The way the brain works is much more complex than simply adding new material to old. Thinking refers to how we see the world, approach problems, and use the different parts of our brain. As science has rapidly improved its ability to measure the brain's activity, our understanding of brain functions has dramatically changed. As research continues, our understanding of this wondrously complex organ is continuing to evolve.

Models have been developed to try to categorize how different individuals think. Early models focused on the specialization of the right and left hemispheres of the brain. Further research has unlocked other complexities and required the development of more complex models. A common current model was developed by Ned Herrmann and shows the brain as four quadrants. Originally, these were four physiological components of the brain. As research continued, it was determined that brain function was too complex to be precisely represented in this format. However, the model has proven effective as a representation of our preferred thinking style and the four quadrants are currently used as a metaphorical model of how we prefer to process information. Currently, the four quadrants are used as a metaphorical model.

Figure 4.1 shows the evolution of the quadrant model from the hemispherical models. Quadrants A and B represent the "left brain" and quadrants C and D represent the "right brain." Herrmann's complete model is detailed in his book *The Creative Brain*. These quadrants illustrate personal preferences and a dominant thinking style. However, most people exhibit characteristics of each quadrant to some degree. Successful people are aware of their strong areas and can capitalize on them while compensating for their weaker areas, and possibly building them up.

The most successful students pick and choose a variety of strategies that work best for their learning style and specific classes. Observe what other successful students are doing and learn from their example. Remember, your one-of-a-kind brain is very capable of learning new material. Spend some time learning how you learn best and you've paved the way for successful learning.

The model can also be used to help understand professors who have a different thinking style. Most people teach like they would like to be taught. If you are in a class where

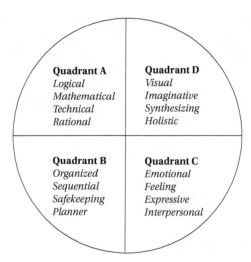

Figure 4.1 Herrmann's "whole brain model" describing four quadrants of the brain (Copyright Herrmann International).

your professor has a very different style than you do, use the model to problem-solve ways to get past this hurdle. It is great experience for later in life. In your professional career, you will have to deal with all different styles effectively to be successful. Being aware of the differences and being able to work with them will be a tremendous asset.

If you would like to know which quadrants apply to you, you can take an assessment profile called the Herrmann Brain Dominance Instrument. An academic advisor or testing center on campus should be able to provide you with information about the HBDI. You may also contact Herrmann International at 794 Buffalo Creek Road, Lake Lure, North Carolina 28746, or at their website at *www.hbdi.com.*

4.8 **Well-Rounded Equals Effective**

Being an effective person goes beyond being a good student. Developing the habits that make you truly effective during your college years will set you up for success in life. Part of being effective is functioning at full capacity. To do this, you must have the various dimensions of your life in order. One analogy that is frequently used is to look at a person as a wheel. A wheel will not roll if it is flat on one side. Similarly, people cannot function optimally if there is a problem area in their life. Five key areas are graphically represented in Figure 4.2. It is important to maintain each area to become as effective as you can be.

Intellectual

One of the main purposes of going to college is to expand the intellectual dimension of your life. Take full advantage of the opportunities at your institution. Besides engineering courses, schools will require general education courses (languages, humanities, or social science classes). These non-technical classes are a great way to broaden yourself. Many students find these classes a refreshing break from their engineering classes.

In addition to course work, colleges have a wide range of activities to enhance your intellectual development. These range from student organizations to seminars to special events. Planning some non-engineering intellectual activities will help to keep you motivated and feeling fresh for your engineering classes. Some students find that reading a novel or two during the semester also provides a refreshing, intellectual break from engineering studies.

Social

The social aspect of college is one that most students seem to master. Planning appropriate social activities will make your college experience more fun. There is a strong correlation between enjoying your college experience and being successful. Also, establishing a group of friends to socialize with will provide a support structure to help you when times get tough. College friends also provide a network that can be very beneficial later in your professional life.

If you have trouble identifying appropriate social activities, there are campus resources to help you. The first place to go is your residence hall staff. Also, most campuses have a central office for student activities. This office can provide you with a list of student organizations and activities.

Physical

All the studying and preparation is a waste of time if you are not physically able to perform when the tests or quizzes come. Also, you can waste time if you are not able to study efficiently. There are three physical areas that are essential to becoming an effective student. They are fitness, sleep, and nutrition.

People are more productive when they are physically fit, and

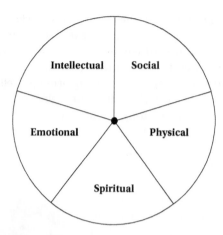

Figure 4.2 Areas of wellness for students.

students are no exception. Exercise is an activity which will not only help you study better but will allow you to live a longer, more productive life. College is a great time to develop a fitness habit. Fitness activities can include aerobics, jogging, biking, walking, or participating in intramural sports. On every campus there are people who can help you develop a fitness plan that is right for you.

Sleep is an area that many students abuse. Finally they are away from their parents and can go to bed any time they want. The problem is that classes still come at the same time regardless of when you fall asleep. Sleep deprivation reduces a person's productivity, which includes studying. Each person needs a different amount of sleep. Typically, most adults need six to eight hours of sleep. It is important to schedule enough time for sleep or you won't be able to be effective in the classroom. Studies have shown that student performance is actually reduced by staying up very late the night before a test to study. You cover more material the night before the test but are so tired at test time that you can't retrieve the material. Honestly evaluate how much sleep you need. Do you fall asleep in class? If so, you probably aren't getting enough rest. Another test for sleep deprivation is to not set an alarm one morning and see when you wake up. If you are on a schedule, most people will wake up at the same time without the alarm. If you sleep several hours past when the alarm would have gone off, you very possibly are sleep-deprived and need more rest.

The third physical area is diet. Again, Mom and Dad are not there to make you eat that broccoli, so you don't. Our bodies are designed to work properly with the right input. While you don't have to become a diet fanatic, eating a sensible diet will enhance your ability to succeed in your academics.

A final note on the physical dimension pertains to non-dietary substances, such as alcohol. The number one problem on most college campuses is alcohol abuse. Every year, thousands of very capable students fail due to alcohol. If you choose to indulge, make honest assessments of the impact on your studying. Don't let alcohol become the barrier to your success.

Spiritual

Few would disagree that humans are spiritual beings and function much better and more efficiently with a balanced spiritual component in their lives. Stephen Covey, in his book *The Seven Habits of Highly Effective People*, puts it this way:

> *The spiritual dimension is your core, your center, your commitment to your value system. It's a very private area of life and a supremely important one. It draws upon the sources that inspire and uplift you and tie you to the timeless truths of all humanity. And people do it very, very differently.*
>
> *I find renewal in daily prayerful meditation on the Scriptures because they represent my value system. As I read and meditate, I feel renewed, strengthened, centered and recommitted to serve.*

Search for the spiritual activities that keep you renewed. This may be through a small group study, attending campus worship services, meditating, or immersing yourself in good music for a time of personal reflection. If you aren't sure where to start, try something which is close to what you were used to before college. This will probably be the most comfortable for you. If it doesn't seem to be the right thing for you, branch out from there.

Emotional

Bill Hybels, pastor of a suburban Chicago church, explained the need to attend to this dimension when he addressed a gathering in 1996 in Detroit. He is an avid runner, eats well, and is physically sound. He has many casual friends and is socially active. He has written a number of books and articles and is sound intellectually. Yet he reached an emotional crisis that prevented him from being effective in his job and with his family.

Hybels describes emotional reserves like a gas tank. We need to put reserves into our tank so we have the ability to take it out when we need it. There will be times in your college career, and beyond, when you need those reserves. The lesson Hybels shared is the need to watch the gas gauge and not to let it run out. Students are no different.

A positive step to take to keep the emotional dimension in check is to set up a support network. This network is what you use to make deposits into your emotional tank. Friends are an integral part of the network. Do you have a person or people whom you can talk honestly with about important issues in your life? This might be a close friend, a parent, a relative, an advisor, a professor, or a counselor. Cultivate at least one friendship where you feel the freedom to share honestly.

The second aspect to maintaining emotional reserves is to schedule time that is emotionally neutral or energizing. Many students get themselves into a lot of activities that are emotionally draining and don't schedule time for themselves for recovery. Eventually this catches up with them and adversely impacts their effectiveness.

A final note on this area is that on each campus there are advisors and counselors who are available to talk to you. Don't feel that you have to have a severe psychological problem to simply talk with them. These people are trained to help you and will just listen if that is what you need. They are a great resource for maintaining the emotional dimension and thereby helping to keep your academic effectiveness high.

4.9 Your Effective Use of Time

You may be wondering how to find time to maintain all these areas every day: studying two hours for each hour spent in class, keeping a balanced lifestyle, etc. And what happened to all that fun time you've heard about? Aren't the college years supposed to be the best years of your life?

The answer is effective time management. Time management is a skill that will help you succeed as a student and will serve you well throughout your entire career. Mastering time management will set you ahead of your peers. It is interesting that when engineers and managers are surveyed and asked why they don't take the time to manage their time, the most common response is inevitably that they don't have time. Students have the same feelings.

If we consider the academic requirements with a 15-credit load, we will have 45 hours per week of class and study time. That could be done Monday through Friday from 8 a.m. until 6 p.m., assuming you take an hour for lunch every day. Think about it. That would give you every evening and every weekend free to have fun or to do other activities.

Many students don't believe they can do this, but the hours add up. The key to making such a schedule work is to make every minute count. Treat school like a full-time job, working similar hours. This will not only help you to effectively manage your life during your academic years, but it also will provide excellent preparation for your future work life.

Let's assume that you have a schedule like the one in Table 4.1. This student has Chemistry, Math, English, Computer Science, and an Introduction to Engineering course. On Monday, the day starts at 8 a.m. with two classes back to back. Then there is an hour break before Computer Science. If you are on the job, you need to be working, and we'll define working as attending class or studying. You need to find a place to study during this hour break. This is where many students go astray. They will waste an hour like this between classes. Similarly, there is a two-hour time block before Math class. This is time to grab lunch and get some more studying done.

It will take some planning to make your day effective because you will need to find suitable places to study. You can use these times to do your individual studying or meet with a study partner or group. It is much more efficient to decide in advance what you will study and when. It can be a waste of precious time to try decide which subject you will study during breaks between classes. Plan enough time for each subject so you can study from class to class and not fall behind in any way. For example, you can't do all your studying for Chemistry on Monday, Math on Tuesday, and Computers on Wednesday or you will fall behind the lectures.

Another thing that scares many students away from effective time management is that it sounds too rigid. One of the reasons for going to college is to have fun and meet people. How can you "go with the flow" if everything is scheduled? The answer is responsible flexibility. If you have your schedule planned out, you can decide when it's a good time to interrupt your "work time" to have some fun. A criterion for making this decision is recovery time. In your schedule, is there enough non-work time to convert to work time before deadlines arrive in your classes? For instance, taking a road trip the day before an exam would be irresponsible. However, right after you have midterms, there might be a window of time to take such a break and still recover.

Scheduling two hours of study per class period is usually enough to keep up during your first year. However, there are classes or times in a semester when this won't be enough. Once you fall behind, it is hard to catch up. To avoid this, a weekly make-up time is an excellent idea. This is like an overtime period at work. At an aerospace company where I worked, Saturday morning from 6 to 10 was my favorite time to schedule this overtime work. This didn't interfere with anything else in my normal

Table 4.1 Sample Class Schedule

	Mon	Tues	Wed	Thurs	Fri
8:00	English		English		English
9:00	Chem 101		Chem 101		Chem 101
10:00		Chem Lab		CS Lab	
11:00	CS 101	Chem Lab	CS 101	CS Lab	
12:00		Chem Lab			
1:00				Chem Rec	
2:00	Math	Math	Math		Math
3:00		Engr 101		Engr 101	
4:00					
5:00					
6:00					
7:00					
8:00					
9:00					

schedule and I could get the work done and still have a full Saturday to enjoy. For some reason, 6 a.m. on Saturday isn't a very popular time for students. However, an evening or two or Saturday afternoon will work, too. These extra time blocks will allow you to keep up and give you the flexibility to take time off with a friend for something fun but still stay on your weekly schedule.

Scheduling Work and Other Activities

No matter how good you get at squeezing in study time between classes, there are still only 24 hours in each day. As a result, there is a limit to what anyone can do in one week. Planning a schedule complete with classes, study time (including make-up time), eating, sleeping, and other activities is essential to deciding how much is too much. If you need to work while you're in school, schedule it into your week. If you join a student organization, plan those hours into your schedule. You need to see where it all fits. A common pitfall with students is that they over-commit themselves with classes, work, and activities. Something has to give. It is easier to drop an activity than to have to repeat a class.

Ray Landis, from California State University at Los Angeles, describes a 60-hour rule to help students decide how much to work. Fifteen credit-hours should equate to 45 hours of effort. That leaves 15 hours for work, totaling 60 hours. Students may be able to work more than this, but each individual has a limit. You need to learn for yourself what your limit is.

If you need to work to make it through college financially, you may need to look at the number of credits you can handle effectively each semester. A schedule can help guide you in determining what is manageable. Before deciding to take fewer credits so you can work more hours, examine the value of the money you can make at your job compared with the starting salary you would expect by graduating a semester earlier. In some cases, it will make sense to work more while taking fewer classes. In others, it will make financial sense to take out a loan, graduate sooner, and pay the loan back with your higher salary.

Calendars

It is important to have a daily and weekly calendar. Another critical item in effective time management is a long-range schedule. For students, this is a semester calendar. This will help you organize for the crunch times, like when all five of your classes have midterms within two days, or you have three semester-projects due on the same day at the end of the semester. It is a fact of student life that tests and projects tend to be bunched together. That is because there are educationally sound times to schedule these things, and most professors follow the same model. Given this fact, you can plan your work schedule so that you spread out your work in a manageable way. This is the part you have control over.

Once, while I was in school, we complained to a professor about this on the first day of class. He offered to give the final exam in the next class period and have the semester project due the second week of the semester to free us up for our other classes. We declined his offer and decided it was best to learn the material first.

Organizers: Paper or Electronic?

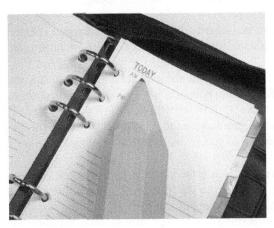

Calendars and task lists can be kept in numerous forms, and you will need to determine which is the most effective for you. There has been a proliferation of electronic tools and devices that are designed to help people communicate and be organized. BlackBerries and cell phones can store tasks and calendars and be used to replace paper calendars. Laptops can be equipped with software such as Microsoft Outlook or Franklin Covey software to manage your tasks and keep appointments. Even with all the technology available, many people still find a role for paper lists or traditional calendars. Just like with the learning styles and study strategies, the method you use to organize your tasks needs to work for you. It does not have to impress your friends with how "tech-savvy" you are.

Do not feel like you have to adopt the latest technology to be effective. The media you use to organize your schedule is a personal choice and needs to work for you. It can be electronic, paper, or a combination, a little of each. The important thing is that it works for you. Our recommendation is to explore different approaches and test how they work and fit for you, your style, and needs. Stay flexible as your needs change and as new approaches and devices come on-line.

Organizing Tasks

A common method for organizing one's day is to make a "To Do" list each morning (or the evening before). This is a good way to ensure that everything needing attention gets it. Things not done on one day's list get transferred over to the next.

The one drawback to making a list is determining what to do first. You can approach the tasks in the order you write them down, but this ignores each task's priority. For instance, if you have a project due the following day for a class and it gets listed last, will it get done? A popular planning tool is the Franklin Planner produced by the Franklin

Covey Co. (Material here used courtesy of Franklin Covey.) They stress the importance of prioritizing your life activities according to your values, personal mission statement, and goals. They suggest using a two-tier approach to prioritizing activities. The first is to use A, B, and C to categorize items as follows:

A = Vital (needs attention today)
B = Important (should be taken care of soon)
C = Optional (no one will notice if it is not done today)

Keep your long-term and intermediate goals in mind when assigning priority. And be sure to write simple statements for each task on your list, beginning each item with an action verb.

Once you have categorized the items, look at all the A priorities. Select the item to do first and give it a 1. This item now becomes A1. The next becomes A2, etc. Now you can attack the items on your list starting with A1 and working through the A's. Once you have finished with the A's, you move on to B1 and work through the B's. Often, you will not get to the C priorities, but you have already decided that it is okay if they don't get done that day.

Table 4.2 shows a sample list for a typical student's day. If you are starting a time management routine, include your list. This keeps you reminded of it, and it gives you something to check off your list right off the bat!

Some lessons can be learned by looking at how this student has scheduled activities. One is that paying the phone bill is not necessarily a top priority, but it requires little time to address. A helpful approach is to look at the quickly done A items and get them out of the way first. Then move on to the more time-consuming items. Also note that a workout was planned into the schedule, right after the chemistry lab. The workout is a priority that comes after some extensive studying. This can help motivate the student to finish a difficult task and then move on to something enjoyable.

Looking at this schedule, you may think that poor Stacy is never going to get called or the letter to the parents written. It often happens that the more fun things are, the more likely they end up a C priority. But it is okay to put relationship-building or fun activities high in priority. People need to stay balanced. Maybe today Stacy is a C, but tomorrow might have that call as an A priority.

The Franklin Covey people provide additional help for identifying priorities. They rate tasks by two criteria, urgency and importance, as shown in Table 4.3. Franklin Covey stresses that most people spend their time in Quadrant I and III. To achieve your long-term goals, Quadrant II is the most important. By spending time on activities that are in II, you increase your capacity to accomplish what matters most to you. The urgency is something that is usually imposed by external forces, whereas importance is something you assign.

You may well want to use a schedule that shows a whole week at a time and has your goals and priorities listed. A weekly schedule allows you to see where each day fits. It also keeps your priorities in full view as you plan. Franklin Covey offers many versions of their Panner, including the Collegiate Edition. Contact them at 1-800-654-1776 or visit their website at www.franklincovey.com.

Table 4.2 Sample Prioritized List of Tasks

Priority	Task
A3	Finish chemistry lab (due tomorrow)
A5	Read math assignment
B3	Do math homework problems
B4	Outline English paper
A4	Workout at Intramural Building
B2	Debug computer program
A6	Annotate math notes
B1	Annotate chemistry notes
A7	Spiritual meditation
C1	Call Stacy
C2	Write parents
A2	Mail phone bill
C3	Shop for CDs
A1	Plan day

Table 4.3 Franklin Covey's 'Time Matrix' for Daily Tasks

	Urgent	Not Urgent
Important	I • Crises • Pressing deadlines • Deadline-driven projects, meetings, preparations	II • Preparation • Prevention • Values clarification • Planning • Relationship building • Recreation • Empowerment
Not Important	III • Interruptions, some phone calls • Some mail, some reports • Some meetings • Many pressing matters • Many popular activities	IV • Trivia, busywork • Some phone calls • Time-wasters • 'Escape' activities • Irrelevant mail • Excessive TV

Copyright 1999 Franklin Covey Co.

Stay Effective

There are two final notes on time management that can help you stay effective. The first is to plan to tackle important activities when you are at your best. We are all born with different biological clocks. Some people wake up first thing in the morning, bright-eyed and ready to work. These people are often not very effective in the late evening. Others would be dangerous if they ever had to operate heavy equipment first thing in the morning, especially before that first cup of coffee. However, once the sun goes down, they can be on a tear and get a lot done. Know yourself well enough to plan most of your studying when you are at your best. It is a common tendency among students to leave studying until late in the evening. This is often because they haven't done it earlier in the day. There are always stories about how the best papers are written at 3:00 a.m. Some are, but most are not. Know yourself well enough to plan effectively.

And finally, pace yourself. You are entering a marathon, not a sprint. Plan your study time in manageable time blocks and plan built-in breaks. A general suggestion is to take a 4–10 minute break for every hour spent studying. This will help keep you fresh and effective.

4.10   **Accountability**

The single most important concept to being successful is to establish accountability for your goals and intentions. The most effective way to do this is with an accountability partner. This is someone whom you trust and from whom you feel comfortable receiving honest feedback. Ideally, you will meet regularly with this person and share how things are going. Regularly means about once per week. Throughout a semester you can fall behind quickly, so weekly checks are a good idea. The person you choose can be a peer, a mentor, a parent, a relative, or any one you feel comfortable with, respect, and trust.

Your accountability could also be to a group of people. Bill Hybels describes the change in his life when he entered into an accountability group with three other men. These men held each other accountable in all areas of their lives—professional, familial, and spiritual. You can do the same as a student.

The first step is to give your accountability partners the standard to which you want to be held. An easy way to do this is to share your goals with them. Let them see and hear from you where you want to be in five years, in ten years, and beyond that. Let them also see the short-term goals you have set to achieve those longer-term objectives. Then, weekly, share with them how you are doing in meeting your goals. You will find that it is much easier to stay on course with the help of at least one close friend. Success will come much easier.

4.11   **Overcoming Challenges**

We all at some point in our lives encounter adversity. As a student, adversity may take the form of a bad grade on a test or for an entire course, a professor you feel is treating you unfairly, not getting into the academic program of your choice, or any of a myriad other college-related obstacles. Since everyone encounters adversity and failure at some point in their life, your personal success is not always dependent on how well you avoid failing, but rather on how you learn from it, move on, and improve. This is especially true as a student.

Almost every student will run into a course or a test where they struggle. The key is to avoid focusing on the negative and continuing to be productive. Every academic advisor has stories about good students who tripped up on a test or in one class and let that affect their performance in other classes.

Focus on "Controllables"

If something bad happens, figure out what can be done. This may mean writing down a list of actions you can take. Don't focus on things you can't change. If you receive a low grade on a test, you can't change the grade. You may, however, be able to talk to the professor about the test. You can study the material you missed and be ready for the next one.

Keep It in Perspective

Examine the real impact of the situation. What are the consequences of the situation? One class can affect your graduation GPA by about 0.05. This will not produce life-altering consequences, unless you would have graduated with a 4.0! If you are unsure about the consequences, see your professor or an academic advisor and have them help you put it in perspective.

Move to the Positive

It is very important if you hit a bump in the road to move beyond it quickly and remain effective in your studies. It is necessary to examine the problem to learn from it. Perhaps you failed a test. This is a great opportunity to examine how you are studying and take corrective action to improve your effectiveness. You can affect your present and future, but not your past.

Remember, Everyone has Setbacks

Again, keep things in perspective. If you run into difficulty, you are not alone. There are numerous examples of setbacks and failures.

Henry Ford neglected to put a reverse gear in his first automobile. But he recovered and went on to be successful.

Abraham Lincoln, likely our greatest president, provided a tremendous example of overcoming adversity. Here is a brief list of some of his accomplishments and setbacks:

1831 – failed in business	1843 – defeated for Congress
1832 – defeated for legislature	1846 – elected to Congress
1833 – failed in business again	1848 – defeated for Congress
1834 – elected to legislature	1855 – defeated for Senate
1835 – fiancée died	1856 – defeated for Vice-President
1838 – defeated for Speaker	1858 – defeated for Senate
1840 – defeated for Elector	1860 – elected President

Lincoln never stopped achieving just because he ran into a setback. You don't have to either.

REFERENCES

Douglass, S. and A. Janssen, *How to Get Better Grades and Have More Fun*, Success Factors, Bloomington, Indiana, 1997.

Maxwell, J. C., *The Winning Attitude, Your Key to Personal Success*, Thomas Nelson Publishers, Nashville, Tennessee, 1993.

Landis, R., *Studying Engineering, A Road Map to a Rewarding Career*, Discovery Press, Burbank, California, 1995.

Tobias, C.U., *The Way They Learn, How to Discover and Teach to Your Child's Strengths*, Focus on the Family, Colorado Springs, Colorado, 1994.

Lumsdaine, E., and M. Lumsdaine, *Creative Problem Solving, Thinking Skills for a Changing World*, 2nd ed., McGraw-Hill, New York, 1993

Herrmann, N., *The Creative Brain*, Brain Books, Lake Lure, North Carolina, 1988.

Covey, S.R., *The 7 Habits of Highly Effective People*, Simon and Schuster, New York, 1989.

EXERCISES AND ACTIVITIES

4.1   Identify three things you can do to keep a positive attitude toward your studies.

4.2   What could you have done if you were in the Physics class as described in Section 4.2?

4.3   Actually writing goals as described is an important element in defining your goals.

   a) Write five goals to achieve by graduation.

   b) Write five goals to reach within ten years.

   c) Write ten goals to reach during your lifetime.

4.4   Write some specific goals:

    a) Set a grade goal for each class this semester.
    b) Calculate your expected semester GPA.
    c) Define a stretch goal for your semester GPA.

4.5   Find out what the following grade point averages are:

    a) Minimum GPA to stay off academic probation
    b) Minimum GPA to be admitted to Tau Beta Pi
    c) Minimum GPA to be admitted to the engineering honor society in your engineering discipline (e.g., Pi Tau Sigma for mechanical engineering, Eta Kappa Nu for electrical engineering)
    d) Minimum GPA to be on the Dean's List or Honors List
    e) Minimum GPA to graduate with honors

4.6   Identify a particular employer and an entry-level position with that employer that you would like to obtain following graduation. Find out what is required to be considered for that position.

4.7   List the barriers you see to achieving your graduation goals.

    a) Categorize them as those within your control and those beyond your control.
    b) Develop a plan to address those within your control.
    c) Develop strategies to cope with those beyond your control.

4.8   For each class you are currently taking:

    a) Map out where you sit in class.
    b) Rate your learning effectiveness (A, B, C, D, or F) for each class.
    c) For each class not rated an A, develop a plan to increase your learning effectiveness.

4.9   For one week, make a log of everything you do. Break it up into categories of attending class, studying, working, playing, eating, sleeping, physical activity, intellectual pursuits, emotional health, spiritual well-being, and other.

    a) Total your time spent on each area. Are you spending the time needed on your studies (2 hours for each hour in class)?
    b) Is this the kind of schedule that will make you successful? If not, what will you change?

4.10   Review your current class schedule and identify times and places you can study between classes.

4.11   Develop a "work" schedule. Identify the times you are "on the job," that is, studying and taking classes. Identify a make-up time for weekly catch-up.

4.12   List all your current courses. Put a "+" next to each one in which you are caught up. Put a "–" next to each one in which you are not caught up right now. ("Caught up" is defined as prepared for the next lecture with all assignments completed and all assigned reading done.)

4.13   Ask one of your professors the main objective of his or her course and report the response. Does this information help you in organizing your studying?

4.14   Annotate your notes from a lecture in one of your courses, then take those notes in to your professor to get his or her comments on your effectiveness. Report the professor's comments.

4.15   Make a list of the students you already know who are pursuing the same major as you. Then make a list of the students you know in your science and math classes. Rate each one according to their suitability as a study partner.

4.16   Identify one or more people with whom you might form a study group. Try studying together on a regular basis until the next test. Report on your experience.

4.17   List the group study rules you used in your group studying. Report on the things that went well and those that did not.

4.18   Make a list of problems you can work, in addition to the assigned homework problems, in one of your math, science, or engineering courses. Report on the method you used to identify these problems.

4.19   Visit one of your professors and report three personal things about that professor. (For example: Where did she go to school? Why did he choose the field he is in? What does she like best about her job? What one piece of advice does he have for freshmen that he didn't share in class?)

4.20   After completing Exercise 19 above, did this change your feelings about your professor? Does he or she seem more approachable now? Did you learn anything else during your visit that could help you succeed in that class?

4.21   Identify at least one opportunity for you as an undergraduate to become involved in engineering research. What advantages could you gain from this experience?

4.22   Identify the type of environment in which you study the best. Describe it and include a sketch of the location.

4.23   Test your conclusion in Exercise 22 with an experiment. Do one assignment in one environment (perhaps in your room, alone and totally quiet) and another in a different environment (perhaps in a more public place, with background noise). Report your findings.

4.24 At the halfway point in the semester, calculate your grades in each class. Compare these to the goals set in Exercise 4. Are you on track? If not, identify the changes you will make to reach these goals.

4.25 Define at least one long-term goal in each of the five wellness areas discussed in this chapter. Next, define short-term goals to achieve these long-term goals.

4.26 List at least two needs you have in each of the five wellness areas. List at least two activities you can plan to meet these needs.

4.27 Rate your roundness in the wellness wheel (Fig. 4.3), using 5 for very well, 4 for well, 3 for okay, 2 for some improvement needed, and 1 for much improvement needed. Do you have a well functioning wheel? If not, identify steps to correct the problem areas. List at least two new activities for each problem area.

4.28 Write a paragraph on how each wellness area can help you succeed:

a) as an engineering student.
b) as a practicing engineer.

4.29 Make a weekly schedule with two hours of studying for every hour spent in class. Identify activities which will help develop the different areas in your life. Is this a sustainable schedule? If the answer is no, identify things which can be dropped.

4.30 Identify on your weekly schedule the times of each day when you study most efficiently. Are these the times you are actually scheduled to study? If not, explain why you've scheduled other items instead of studying at these times.

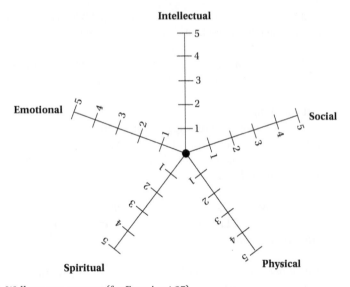

Figure 4.3 Wellness assessment (for Exercise 4.27).

4.31   Adopt one of the weekly schedule instruments discussed in this chapter. Try using it for two weeks and report the results. Was this the right instrument for you?

4.32   Make a "To Do" list. Prioritize the list using A, B and C categories, and then number the items in each category.

4.33   Place the items of Exercise 4.32 into the four quadrants of Covey's matrix shown in Table 4.3. How many A priorities are in quadrant II? Based on this ranking of your "To Do" list, would you change any of your priorities?

4.34   List the top three things you are failing to do that would most help you succeed in your career. Briefly explain why you are currently failing to do each item.

4.35   List three people you would consider for an accountability partner.

4.36   Make a list of at least five questions you want your accountability partner to ask you at each check-up.

4.37   Meet with an accountability partner for one month. Report back on your experience. Did it help? Is this something you will continue?

4.38   Call home and tell your parents that you are developing a success strategy for college. Explain the strategy to them.

4.39   Assume that your grades will be lower than expected at the end of the semester, and write out a strategy for recovering from these low grades.

4.40   Make a list of at least three university people who can help you deal with a problem that is preventing you from succeeding. List each by name and position.

4.41   Make a list of your activities over the last 24 hours. Compare the list and the hours spent on each with your planned schedule.

4.42   List your long-term goals, including every area of your life. Fill out a schedule of what you actually did during the last seven days. Put each activity into categories corresponding to one of your long-term goals, or an extra category titled "not related." Grade yourself (using A, B, C, D, or F), indicating how well you did the past week working toward your long-term goals.

4.43   Take the inventory test discussed in Section 4.7 on memory languages. Which is your dominant mode? List three actions you can take to improve your learning based on this dominant mode.

4.44   Which of Herrmann's quadrants best describes your dominant thinking style? Which one is the weakest?

4.45   Which of Herrmann's four quadrants is best suited for engineering? Are there any dominant modes of thinking that are not ideal for engineering? Explain your answers.

4.46   List all your current professors and identify their probable Herrmann quadrant. Compare this with your own style. For the professors with styles that differ from yours, develop an action plan to deal with these differences.

4.47   Explore the availability of electronic organizers and prepare a brief report on the options for students.

4.48   Compare and contrast electronic organizers versus paper ones. What are the benefits of each and which would you recommend to your classmates?

4.49   Interview two professors and report what they use to organize and prioritize their time.

CHAPTER 5

Problem Solving

5.1 Introduction

Seven-year-old Dan could build all kinds of things: toy cars, ramps, and simple mechanical gadgets. Even more amazing was that he could do all this with four simple tools—scissors, hammer, screwdriver, and duct tape. One Christmas, his father asked if he'd like more tools. "No thanks, Dad" was his response. "If I can't make it with these, it's not worth making." It wasn't long, however, before Dan realized that the cool stuff he could create or the problems he could solve were dependent on the tools he had. So he began gathering more tools and learning how to use them. Today, he

still solves problems and makes a lot of cool stuff—like real cars, furniture, playgrounds, and even houses. His tools fill a van, a workshop, and a garage. Having the right tools and knowing how to use them are critical components of his success.

Likewise, most people, when they are children, learn a few reliable ways to solve problems. Most fifth graders in math class know common problem-solving methods like "draw a picture," "work backwards," or the ever-popular "guess and check" (check in the back of the textbook for the correct answer, that is). Unfortunately, many people choose to stop adding new problem-solving tools to their mental toolbox. "If I don't know how to solve it, it's probably not worth solving anyway," they may think. Or, "I'll pass this problem over for now and get back to my usual tasks." This severely limits them. Engineers, by definition, need to be good at solving problems and making things. Therefore, filling your toolbox with the right tools and knowing how to use them are critical components to becoming a successful engineer.

The goal of this chapter is to increase the number of problem-solving strategies and techniques you commonly use, and to enhance your ability to apply them creatively in various problem-solving settings. Hopefully, this will be only the beginning of a lifelong process of gaining new mental tools and learning how to use them.

5.2   **Analytic and Creative Problem Solving**

The toolbox analogy is very appropriate for engineering students. Practicing engineers are employed to solve a very wide variety of problems. In daily practice, engineers can end up filling a number of different roles in daily practice. These roles can require very different problem-solving abilities. Problem solving can be broken into Analytic and Creative. Most students are more familiar with Analytic problem solving where there is one correct answer. In Creative problem solving there is no single right answer. Your analytic tools represent what's in your toolbox. Your creative skills represent how you handle your tools.

To better understand the difference between these two kinds of problem solving, let's examine the one function that is common to all engineers: design. In the Design chapter, we detail a design process with ten steps. Those ten steps are:

1. Identify the problem
2. Define the working criteria/goal(s)
3. Research and gather data
4. Brainstorm for creative ideas
5. Analyze
6. Develop models and test
7. Make the decision
8. Communicate and specify
9. Implement and commercialize
10. Prepare post-implementation review and assessment

By definition, design is open-ended and has many different solutions. The automobile is a good example. Sometime when you are riding in a car, count the number of different designs you see while on the road. All of these designs were the result of engineers solving a series of problems involved in producing an automobile. The design process as a whole is a **creative problem-solving process**.

The other method, **analytic problem solving**, is also part of the design process. Step 5, Analyze, requires analytic problem solving. When analyzing a design concept, there is only one answer to the question "Will it fail?" A civil engineer designing a bridge might employ a host of design concepts, though at some point he or she must accurately assess the loads the bridge can support.

Most engineering curricula provide many opportunities to develop analytic problem-solving abilities, especially in the first few years. This is an important skill. Improper engineering decisions can put the public at risk. For this reason, it is imperative that proper analytical skills be developed.

As an engineering student you will be learning math, science, and computer skills that will allow you to tackle very complex problems later in your career. These are critical, and are part of the problem-solving skill set for analysis. However, other tools are equally necessary.

Engineers also can find that technical solutions applied correctly can become outmoded, or prove dangerous, if they aren't applied with the right judgment. Our creative skills can help greatly in the big-picture evaluation, which can determine the long-term success of a solution. The solution of one problem can cause others if foresight is neglected, and this is not easily avoided in all cases. For instance, the guilt that the developers of dynamite and atomic fission have publicly expressed relates to a regret for lack of foresight. Early Greek engineers invented many devices and processes which they refused to disclose because they knew they would be misused. The foresight to see how technically correct solutions can malfunction or be misapplied takes creative skill to develop. What about fertilizers which boost crops but poison water supplies? Modern engineers need to learn how to apply the right solution at the right time in the right way for the right reason with the right customer.

Most people rely on two or three methods to solve problems. If these methods don't yield a successful answer, they become stuck. Truly exceptional problem solvers learn to use multiple problem-solving techniques to find the optimum solution. The following is a list of possible tools or strategies that can help solve simple problems:

1. Look for a pattern
2. Construct a table
3. Consider possibilities systematically
4. Act it out
5. Make a model
6. Make a figure, graph or drawing
7. Work backwards
8. Select appropriate notation
9. Restate the problem in your own words
10. Identify necessary, desired, and given information
11. Write an open-ended sentence
12. Identify a sub-goal
13. First solve a simpler problem
14. Change your point of view
15. Check for hidden assumptions
16. Use a resource
17. Generalize
18. Check the solution; validate it
19. Find another way to solve the problem
20. Find another solution
21. Study the solution process
22. Discuss limitations
23. Get a bigger hammer
24. Sleep on it
25. Brainstorm
26. Involve others

In both analytic and creative problem solving, there are different methods for tackling problems. Have you ever experienced being stuck on a difficult math or science problem? Developing additional tools or methods will allow you to tackle more of these problems effectively, making you a better engineer.

The rest of this chapter is a presentation of different ways of solving problems. These are some of the techniques that have been shown to be effective. Each person has a unique set of talents and will be drawn naturally to certain problem-solving tools. Others will find a different set useful. The important thing as an engineering student is to experiment and find as many useful tools as you can. This will equip you to tackle the wide range of challenges which tomorrow's engineers will face.

5.3 Analytic Problem Solving

Given the importance of proper analysis in engineering and the design process, it is important to develop a disciplined way of approaching engineering problems. Solving analytic problems has been the subject of a great deal of research, which has resulted in several models. One of the most important analytic problem-solving methods that students are exposed to is the Scientific Method. The steps in the Scientific Method are as follows:

1. Define the problem
2. Gather the facts
3. Develop a hypothesis
4. Perform a test
5. Evaluate the results

In the Scientific Method, the steps can be repeated if the desired results are not achieved. The process ends when an acceptable understanding of the phenomenon under study is achieved.

In the analysis of engineering applications, a similar process can be developed to answer problems. The advantage of developing a set method for solving analytic problems is that it provides a discipline to help young engineers when they are presented with larger and more complex problems. Just like a musician practices basic scales in order to set the foundation for complex pieces to be played later, an engineer should develop a sound fundamental way to approach problems. Fortunately, early in your engineering studies you will be taking many science and math courses that will be suited to this methodology.

The Analytic Method we will discuss has six steps:

1. Define the problem and make a problem statement
2. Diagram and describe
3. Apply theory and equations
4. Simplify the assumptions

5. Solve the necessary problems
6. Verify accuracy to required level

Following these steps will help you better understand the problem you are solving and allow you to identify areas where inaccuracies might occur.

Step 1: Problem Statement

It is important to restate the problem you are solving in your own words. In textbook problems, this helps you understand what you need to solve. In real life situations, this helps to ensure that you are solving the correct problem. Write down your summary, then double-check that your impression of the problem is the one that actually matches the original problem. Putting the problem in your own words is also an excellent way to focus on the part of the problem you need to solve. Often, engineering challenges are large and complex, and the critical task is to understand what part of the problem you need to solve.

Step 2: Description

The next step is to describe the problem and list all that is known. In addition to restating the problem, list the information given and what needs to be found. This is shown in Example 5.2, which follows this section. Typically, in textbook problems, all the information given is actually needed for the problem. In real problems, more information is typically available than is needed to do the calculations. In other cases, information may be missing. Formally writing out what you need and what is required helps you to clearly sort this out.

It is also helpful to draw a diagram or sketch of the problem you are solving to be able to understand the problem. Pictures help many people to clarify the problem and what is needed. They are also a great aid when explaining the problem to someone else. The old saying could be restated as "A picture is worth a thousand calculations."

Step 3: Theory

State explicitly the theory or equations needed to solve the problem. It is important that you write this out completely at this step. You will find that most real problems and those you are asked to solve as an undergraduate student will not require exact solutions to complete equations. Understanding the parts of the equations that ought to be neglected is vital to your success.

It is not uncommon to get into a routine of solving a simplified version of equations. These may be fine under certain conditions. An example can be seen in the flow of air over a body, like a car or airplane wing. At low speeds, the density of the air is considered a constant. At higher speeds this is not the case, and if the density were considered a constant, errors would result. Starting with full equations and then simplifying reduces the likelihood that important factors will be overlooked.

Step 4: Simplifying Assumptions

As mentioned above, engineering and scientific applications often cannot be solved precisely. Even if they are solvable, determining the solution might be cost prohibitive. For instance, an exact solution might require a high-speed computer to calculate for a year to get an answer, and this would not be an effective use of resources. Weather prediction at locations of interest is such an example.

To solve the problem presented in a timely and cost-effective manner, simplifying assumptions are required. Simplifying assumptions can make the problem easier to solve and still provide an accurate result. It is important to write the assumptions down along with how they simplify the problem. This documents them and allows the final result to be interpreted in terms of these assumptions.

While estimation and approximation are useful tools, engineers also are concerned with the accuracy and reliability of their results. Approximations are often possible if assumptions are made to simplify the problem at hand. An important concept for engineers to understand in such situations is the Conservative Assumption.

In engineering problem solving, "conservative" has a non-political meaning. A conservative assumption is one that introduces errors on the safe side. We mentioned that estimations can be used to determine the bounds of a solution. An engineer should be able to look at those bounds and determine which end of the spectrum yields the safer solution. By selecting the safer condition, you are assuring that your calculation will result in a responsible conclusion.

Consider the example of a materials engineer selecting a metal alloy for an engine. Stress level and temperature are two of the parameters the engineer must consider. At the early stages of a design, these may not be known. What he or she could do is estimate the parameters. This could be done with a simplified analysis. It could also be done using prior experience. Often, products evolve from earlier designs for which there are data. If such data exist, ask how the new design will affect the parameters you are interested in. Will the stress increase? Will it double? Perhaps a conservative assumption would be to double the stress level of a previous design.

After taking the conservative case, a material can be selected. A question the engineer should ask is: "If a more precise answer were known, could I use a cheaper or lighter material?" If the answer is No, then the simple analysis might be sufficient. If Yes, the engineer could build a case for why a more detailed analysis is justified.

Another example of conservatism can be seen in the design of a swing set (see Fig. 5.1). In most sets, there is a horizontal piece from which the swings hang, and an inverted V-shaped support on each end. If you were the design engineer, you might have to size these supports.

One of the first questions to ask is, "For what weight do we size the swing set?" One way to answer this question is to do a research project on the weight distribution of children, followed by doing research into the use of swings by children at different ages, and then analyzing those data to determine typical weights of children that would use such swings. You might even need to observe playground activity to see if several children pile on a swing or load the set in different ways—say, by climbing on it. This might take weeks to do, but you would have excellent data.

On the other hand, you could just assume a heavy load that would exceed anything the children would produce. For this case, say, 500 pounds per swing. That would assume the equivalent of two large adults on each swing. Make the calculation to determine the needed supports, and then ask if it makes sense to be more detailed.

To answer this, one has to answer the question "What problem am I solving?" Am I after the most precise answer I can get? Am I after a safe and reliable answer regardless of other concerns?

In engineering, the "other concerns" include such things as cost and availability. In the current example, the supports would most likely be made of common materials. Swing sets are typically made of wood or steel tubing. So, one way to answer the above questions is to quickly check and see if making a more detailed analysis would be justified. Check the prices of the materials needed for your conservative assumption against potential materials based on a more detailed analysis. Does the price difference justify a more detailed analysis? Another way to ask this is "Does the price difference justify spending time to do the analysis?" Remember, as an engineer, your time costs money.

The answer may be, "It depends." It would depend on how much the difference is and how many you are going to produce. It wouldn't make sense to spend a week

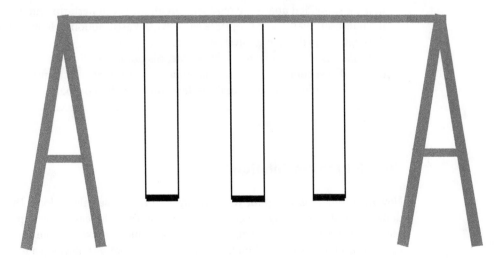

Figure 5.1 Simple swing set.

analyzing or researching the answer if it saves you or your company only $500. If your company was producing a million sets and you could save $5 per set, it would justify that week's work.

Approximation can improve decision-making indirectly as well. By using it to quickly resolve minor aspects of a problem, you can focus on the more pressing aspects.

Engineers are faced with these types of decisions every day in design and analysis. They must accurately determine what problem is to be solved, and how. Safety of the public is of paramount concern in the engineering profession. You might be faced with a short-term safety emergency where you need to quickly prioritize solutions. Or you might be designing a consumer product where you have to take the long-term view. Swing sets are often used for generations, after all.

As an engineer, you will need to develop the ability to answer, "What problem am I really solving?" and, "How do I get the solution I need most efficiently?"

EXAMPLE 5.1 A manager for an aerospace firm had a method for "breaking in" young engineers. Shortly after the new engineer came to work in his group, the manager would look for an appropriate analysis for the young engineer. When one arose, he would ask the young engineer to perform the analysis. The engineer would typically embark on a path to construct a huge computer model requiring lengthy input and output files that would take a few weeks to complete. The manager would return to the engineer's desk the following morning, asking for the results. After hearing the response that all that was really accomplished was to map out the next three weeks worth of work, the manager would erupt and explain at a very high volume that he or she had been hired as an engineer and what they were supposed to be doing was an engineering analysis, not a "science project." He would show the young engineer how doing a simplified analysis would take an afternoon to perform and would answer the question given the appropriate conservative assumptions. He would then leave the engineer to the new analysis method.

While his methods were questionable, his point is well taken. There are always time and cost concerns as well as accuracy. If conservative assumptions are made, safe and reliable design and analysis decisions can often be made.

Step 5: Problem Solution

Now the problem is set up for you actually to perform the calculations. This might be done by hand or by using the computer. It is important to learn how to perform simple hand calculations; however, computer applications make complex and repetitive calculations much easier. When using computer simulations, develop a means to document what you have done in deriving the solution. This will allow you to find errors faster, as well as to show others what you have done.

Step 6: Accuracy Verification

Engineers work on solutions that can affect the livelihood and safety of people. It is important that the solution an engineer develops is accurate.

I had a student once whom I challenged on this very topic. He seemed to take a cavalier attitude about the accuracy of his work. Getting all the problems mostly correct would get him a B or a C in the class. I tried to explain the importance of accuracy and how people's safety might depend on his work. He told me that in class I caught his mistakes, and on the job someone would check his work. This is not necessarily the case! Companies do not hire some engineers to check other engineers' work. In most systems, there are checks in place. However, engineers are responsible for verifying that their own solutions are accurate. Therefore, it is important for the student to develop skills to meet any required standard.

A fascinating aspect of engineering is that the degree of accuracy is a variable under constant consideration. A problem or project may be properly solved within a tenth of an inch, but accuracy to a hundredth might be unneeded and difficult to control, rendering that kind of accuracy incorrect. But the next step of the solution might even require accuracy to the thousandth. Be sure of standards!

There are many different ways to verify a result. Some of these ways are:

- Estimate the answer.
- Simplify the problem and solve the simpler problem. Are the answers consistent?
- Compare with similar solutions. In many cases, other problems were solved similarly to the one currently in question.
- Compare to previous work.
- Ask a more experienced engineer to review the result.
- Compare to published literature on similar problems.
- Ask yourself if it makes sense.
- Compare to your own experience.
- Repeat the calculation.
- Run a computer simulation or model.
- Redo the calculation backwards.

It may be difficult to tell if your answer is exactly correct, but you should be able to tell if it is close or within a factor of 10. This will often flag any systematic error. Being able to back up your confidence in your answer will help you make better decisions on how your results will be used.

The following example will illustrate the analytical method.

EXAMPLE 5.2 Problem: A ball is projected from the top of a 35 m tower at an angle 25° from horizontal with an initial velocity $v_o = 80$ m/s. What is the time it takes to reach the ground, and what is the horizontal distance from the tower to the point of impact?

Problem Statement: Given: Initial velocity $v_0 = 80$ m/s
Initial trajectory $= 25°$ from horizontal
Ball is launched from a height of 35 m

Find: 1) Time to impact
2) Distance of impact from the tower

Equations: Found in a physics text:

(a) $x_i = x_0 + v_0 t \cos \theta + \dfrac{1}{2} a_x t^2$

(b) $y_1 = y_0 + v_0 t \sin \theta + \dfrac{1}{2} a_y t^2$

Assumptions: Neglect air resistance
Acceleration $= -9.81$ is constant in the vertical direction
Acceleration is zero in the horizontal direction
Ground is level and impact occurs at $y = -35$ m

Solution: Begin with equation (b):

$$-35 = 080 t \sin 25° - \frac{1}{2} 9.8 t^2 \text{ or } t^2 - 6.90 t - 7.14 = 0$$

$$\therefore t = \frac{b \pm \sqrt{b^2 - 4ac}}{2a} = \frac{\sqrt{6.9^2 + 4 \times 1 \times 7.14}}{2} = 7.81 \text{ sec}$$

The solution must be 7.81 sec, since time must be positive.
The distance is found using equation (a):

$$x_1 = 0 + 80 \times 7.81 \times \cos 25° = 566 \text{m}$$

Estimation

Estimation is a problem solving tool that engineers need to develop. It can provide quick answers to problems and to verify complicated analyses. Young engineers are invariably amazed when they begin their careers at how the older, more experienced engineers can estimate so closely to problems before the analysis is even done.

I was amazed at one such case that I experienced first hand. I was responsible for doing a detailed temperature analysis on a gas turbine engine that was to go on a new commercial aircraft. We presented our plan to some managers and people from the Chief Engineers Office. The plan called for two engineers working full time for a month on the analysis.

One of the more senior engineers who had attended the presentation sent me some numbers the next day. He had gone back to his office and made some approximate calculations and sent me the results. His note told me to check my results against his numbers.

After the month long analysis, which involved making a detailed computer model of the components, I checked our results with his "back of the envelope" calculations. He was very close to our result! His answers were not exactly the same as ours, but they were very close. The detailed analysis did provide a higher degree of precision. However, the senior engineer was very close.

What he had done was simply use suggestion #13 from the problem-solving methods. He solved a simpler problem. He used simple shapes with known solutions to approximate the behavior of the complex shapes of the jet engine's components.

While estimation or approximation may not yield the precision you require for an engineering analysis, it is a very useful tool. One thing that is can do is to help you check your analysis. Estimation can provide bounds for potential answers. This is especially critical in today's dependence on computer solutions. Your method may be correct, but one character mistyped will throw your results off. It is important to be able to have confidence in your results by developing tools to verify accuracy.

Estimation can also be used to help decide if a detailed analysis is needed. Estimating using a best case and worst case scenario can yield an upper and lower bound to the problem. If the entire range of potential solutions is acceptable, why do the detailed analysis? What if, for our case, the whole range of temperatures met our design constraints? It would not have justified a month of our work. In this case, the added precision was needed and did lead to a design benefit. As engineers, you will be asked to decide when a detailed analysis is required and when it is not.

Estimation can be a very powerful tool to check accuracy and make analysis decision. Senior engineers in industry often comment that current graduates lack the ability to do an approximation. This has become more important with the dependence on computer tools and solutions. Computer analyses lend themselves to typos on inputs or arithmetic errors. Being able to predict the ballpark of the expected results can head off potentially disastrous results.

EXAMPLE 5.3   With all of today's sophisticated analytical tools, it may be hard to imagine that accurate measurements could be made without them. Take the size of a molecule. Today's technology makes this measurement possible. Could you have made this measurement in the 18th century? Benjamin Franklin did.

He found the size of an oil molecule. He took a known volume of oil and poured it slowly into a still pool of water. The oil spread out in a circular pattern. He measured the circle and calculated the surface area. Since the volume of the oil remained the same, he calculated the height of the oil slick and assumed that was the molecular thickness, which turned out to be true. His result was close to the correct value.

Do we have more accurate measurements now? Certainly we do. However, Benjamin Franklin's measurement was certainly better than no value. As engineers, there will be times when no data exists and you will need to make a judgment. Having an approximate result is much better than no data.

5.4   Creative Problem Solving

Many engineering problems are open-ended and complex. Such problems require creative problem solving. To maximize the creative problem-solving process, a systematic approach is recommended. Just as with analytic problem solving, developing a systematic approach to using creativity will pay dividends in better solutions.

Revisiting our toolbox analogy, the creative method provides the solid judgment needed to use your analytic tools. Individual thinking skills are your basics here. The method we'll outline now will help you to apply those skills and choose solution strategies effectively.

There are numerous ways to look at the creative problem-solving process. We will present a method which focuses on answering these five questions:

1. What is wrong?
2. What do we know?
3. What is the real problem?
4. What is the best solution?
5. How do we implement the solution?

By dividing the process into steps, you are more likely to follow a complete and careful problem-solving procedure, and more effective solutions are going to result. This also allows you to break a large, complex problem into simpler problems where your various skills can be used. Using such a strategy, the sky is the limit on the complexity of projects which an engineer can complete. It's astonishing and satisfying to see what can be done.

Divergence and Convergence

At each phase of the process, there is both a divergent and a convergent part of the process, as shown in Figure 5.2. In the divergent process, you start at one point and reach for as many ideas as possible. Quantity is important. Identifying possibilities is the goal of the divergent phase of each step. In this portion of the process (indicated by the outward pointing arrows, $\leftarrow\rightarrow$) use the brainstorming and idea-generating techniques described in subsequent sections. Look for as many possibilities as you can. Often, the best solutions come from different ideas.

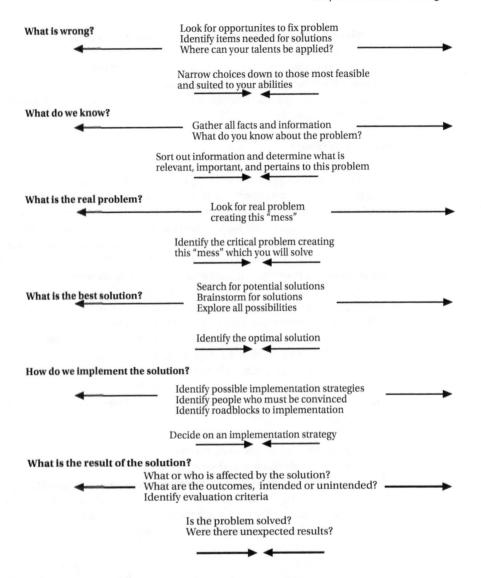

Figure 5.2 Problem-solving process.

In the convergence phase (indicated by inward pointing arrows, →←) use analytical and evaluative tools to narrow the possibilities to the one(s) most likely to yield results. Quality is most important, and finding the best possibility to move the process to the next phase is the goal. (One common method is to use a matrix to rate ideas based on defined criteria.) If one choice fails to produce satisfactory results, go back to the idea lists for another method to tackle the problem.

Sample Problem: Let's Raise Those Low Grades!

In order to illustrate the problem-solving process, let's consider an example of a problem and work through the process. The problem we will use for illustration is a student who is getting low grades. This certainly is a problem that needs a solution. Let's see how the process works.

What's Wrong?

In the first step of the problem-solving process, an issue is identified. This can be something stated for you by a supervisor or a professor, or something you determine on your own. This is the stage where entrepreneurs thrive—looking for an opportunity to meet a need. Similarly, engineers look to find solutions to meet a need. This may involve optimizing a process, improving customer satisfaction, or addressing reliability issues.

To illustrate the process, let's take our example of the initial difficulty that beginning engineering students can have with grades. Most students were proficient in high school and didn't need to study much. The demands of engineering programs can catch some students off guard. Improving grades is the problem we will tackle.

At this stage, we need to identify that there is a problem that is worth our effort in solving. A good start would be to identify whether it truly is a problem. How could you do this? Total your scores thus far in your classes and compare them with the possible scores listed in your syllabus. Talk with your professor(s) to find out where you are in the classes and if grades will be given on a curve. Talk to an advisor about realistic grade expectations. At the end of this process, you should know if you have a grade problem.

EXAMPLE 5.4   Engineers are trained problem solvers. Many engineers move into other careers where their problem-solving abilities are valued. One such case is Greg Smith, who was called into the ministry after becoming a mechanical engineer.

"As soon as people discover that I have a degree in mechanical engineering and am in the ministry, they always ask me the same question, 'So, do you still plan on using your engineering degree some day?' It is difficult for people to understand that I use my engineering training every day as a full time minister.

"One of the greatest skills I learned at Purdue was problem solving. Professors in mechanical engineering were notorious for giving us more information than we would need to solve the problem. Our job was to sift through all the facts and only use what was needed to reach a solution. I find myself using this very skill whenever I am involved in personal counseling. For example, just this week I was listening to someone unload about all the stress he was experiencing from life. He listed his job, a second business at home, family, and finances as the primary sources of stress. After further conversation, I was able to 'solve' his problem by showing that most of his stress was related to only one of the four sources he listed. The root issue was that this person was basing much of his self-image on his success in this one area of life. I helped him regain his sense of balance in seeing his self-worth not in job performance, but in who he was as a whole person."

What Do We Know?

The second step in problem solving is the gathering of facts. All facts and information related to the problem identified in the first step are gathered. In this initial information gathering stage, do not try to evaluate whether the data are central to the problem. As will be explained in the brainstorming and idea generating sections to follow, premature evaluation can be a hindrance to generating adequate information.

With our example of the student questioning his grades, the information we generate might include:

1. Current test grades in each course
2. Current homework and quiz grades in each course
3. Percent of the semester's grade already determined in each course
4. Class average in each course
5. Professors' grading policies
6. Homework assignments behind
7. Current study times
8. Current study places
9. Effective time spent on each course
10. Time spent doing homework in each course
11. Performance of your friends
12. Your high school grades
13. Your high school study routine

Part of this process is to list the facts that you know. To be thorough, you should request assistance from someone with a different point of view. This could be a friend, an advisor, a parent, or a professor. They might come up with a critical factor you have overlooked.

What Is the Real Problem?

This stage is one that is often skipped, but it is critical to effective solutions. The difference between this stage and the first is that this step of the process answers the question *why*. Identifying the initial problem answers the question of *what*, or what is wrong. To effectively fix the problem, a problem solver needs to understand *why* the problem exists. The danger is that only symptoms of the problem will be addressed, rather than root causes.

In our example, we are problem solving low class grades. That is the *what* question. To really understand a problem, the *why* must be answered. Why are the grades low? Answering this question will identify the cause of the problem. The cause is what must be dealt with for a problem to be effectively addressed.

In our case, we must understand why the class grades are low. Let us assume that poor test scores result in the low grades. But this doesn't tell us why the test scores are low. This is a great opportunity for brainstorming potential causes. In this divergent phase of problem definition, don't worry about evaluating the potential causes. Wait until the list is generated.

Possible causes may include:

- Poor test-taking skills
- Incomplete or insufficient notes
- Poor class attendance
- Not understanding the required reading
- Not spending enough time studying
- Studying the wrong material
- Attending the wrong class
- Studying ineffectively (e.g., trying to cram the night before tests)
- Failing to understand the material (*note*: might reveal need for tutor or study group)
- Using solution manuals or friends as a crutch to do homework
- Not at physical peak at test time (i.e., up all night before the test)
- Didn't work enough problems from the book

After you have created the list of potential causes, evaluate each as to its validity. In our example, there may be more than one cause contributing to the low grades. If so, make a rank-ordered list. Rank the causes in order of their impact on class performance.

Assume the original list is reduced to:

- Poor class attendance
- Studying ineffectively
- Not spending enough time studying
- Not understanding the material

Of these things, rank them in order of their impact on grades. You may be able to do this yourself or you may need help. An effective problem solver will seek input from others when it is appropriate. In this case, a professor or academic advisor might be able to help you rank the causes and help determine which ones will provide the greatest impact. For this example, say the rank order was:

1. Not understanding the material
2. Studying ineffectively
3. Not spending enough time studying
4. Poor class attendance

We determined that while class attendance was not perfect, it was not the key factor in the poor performance. The key item was not understanding the material. This goes with ineffective studying and insufficient time.

Identifying key causes of problems is vital, so let's look at another example.

**EXAMPLE
5.5**   **"What was the confusion?"** Defining the real problem is often the most critical stage of the problem-solving process and the one most often skipped. In engineering, this can be the difference between a successful design and a failure.

An example of this occurred with an engine maker. A new model engine was introduced and began selling quickly. A problem developed, however, and failures started to happen. Cracks initiating from a series of bolt holes were causing the engines to fail. A team of engineers was assembled and the part was redesigned. They determined that the cause was high stress in the area of the bolt hole. Embossments were added to strengthen the area. It was a success.

Until, that is, the redesigned engines accumulated enough time in the field to crack again. A new team of engineers looked at the problem and quickly determined that the cause was a three-dimensional coupling of stress concentrations created by the embossments, resulting in too high a stress near the hole where the crack started. The part was redesigned and introduced into the field. It was a success.

Well, it was a success, until the redesigned engines accumulated enough time in the field to crack yet again. A third team of engineers was assembled. This time, they stepped back and asked what the real cause was. Part of the team ran sophisticated stress models of the part, just like the other two teams had done. The new team members looked for other causes. What was found was that the machine creating the hole introduced microcracks into the surface of the bolt holes. The stress models predicted that further stresses would start a crack. However, if a crack were intentionally introduced, the allowable stresses were much lower and the part would not fail. The cause wasn't with cracking but with the wrong kind of cracking. The third team was a success because it had looked for the real problem. Identifying the real cause initially, which in this case involved the machining process, would have saved the company millions of dollars, and spared customers the grief of the engine failures.

What Is the Best Solution?

Once the problem has been defined, potential solutions need to be generated. This can be done by yourself or with the help of friends. In an engineering application, it is wise to confer with experienced experts about the problem's solution. This may be most productive after you have begun a list of causes. The expert can comment on your list and offer his own as well. This is a great way to get your ideas critiqued. After you gain more experience, you will find that technical experts help you narrow down the choices rather than providing more. Also, go to more than one source. This may provide more ideas as well as help with the next step.

In our example, the technical expert may be a professor or advisor who is knowledgeable about studying. Let's assume that the following list of potential solutions has been generated.

- Get a tutor
- Visit the professors during office hours

- Make outside appointments with professors
- Visit help rooms
- Form study group
- Outline books
- Get old exams
- Get old sets of notes
- Outline lecture notes
- Review notes with professors
- Do extra problems
- Get additional references
- Make a time schedule
- Drop classes
- Retake the classes in the summer
- Go to review sessions

The list now must be evaluated and the best solution decided upon. This convergent phase of the problem-solving process is best done with the input of others. It is especially helpful to get the opinions of those who have experience and/or expertise in the area you are investigating. In an engineering application, this may be a lead engineer. In our case, it could be an academic advisor, a teaching assistant, or a professor. These experts can be consulted individually or asked to participate as a group.

When refining solutions, an effective tool is to ask yourself which of them will make the biggest impact and which will require the most effort. Each solution gets ranked high or low as to impact and effort. Ideal solutions are those that have a high impact yet require a low level of effort. A visual tool which helps make this process easier is the problem-solving matrix shown in Figure 5.3. Solutions that fall in quadrant B should be avoided. These solutions require a great deal of effort for marginal payoff. Solutions that are the most effective are those in quadrant C. These are the ones that are easy to implement and will produce significant results. It may be necessary to include solutions in quadrant D, requiring high effort and yielding high impact. These should be carefully evaluated to ensure that the investment is worth the reward.

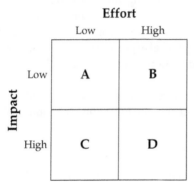

Figure 5.3 A Problem-Solving Matrix.

Implementing the Solution

Implementing the solution is a step that may seem trivial and not worth discussion. This step, however, is a critical phase of the problem-solving process. In our scenario, the solution selected may have been:

- Do extra problems

To accomplish this solution, appropriate additional problems must be selected, done, and corrected. Implementation probably requires the assistance of the instructor to help select appropriate problems. To randomly pick additional practice problems could move the effort from quadrant C to D and even B. To have the greatest chances of success, you would want to gain the assistance of your instructor. Other sources of assistance could come from a study group which poses the challenges of getting them to do extra problems with you.

As with all the other steps, a divergent phase begins the step including such activities as brainstorming. The convergent phase of the step culminates with the selection of the implementation plan.

In engineering applications, implementation can be a very critical phase of the process. Most of the solutions to problems either require additional resources including money or the cooperation of other groups that are either not directly affected by the problem or not under your control. Especially early in your career, you will not have control over the people who need to implement solutions so you will need to sell your ideas to the managers who do. An effective implementation plan will be critical to getting your solutions accomplished.

President Jimmy Carter was educated as an engineer and was one of the most respected presidents as a strong moral person. During his presidency, the country was in an energy crisis and action was needed to address the problems. The story goes that he summoned Tip O'Neal, the speaker of the House of Representatives, to the White House to explain his plan. Mr. O'Neal responded by giving the President a list of people who would need to be sold on the plan before it went public. President Carter who was new to Washington was going to skip the development of an implementation plan and enact the plan. He believed that with the country in such an energy crisis that it was obvious what was needed and cooperation was a given. Tip O'Neal was very experienced in the Washington political system and understood that they needed to spend time gaining acceptance to their plan before it could be implemented.

Engineering situations can be very similar. While most are not as political as Washington D.C., most require gaining acceptance of others to implement a solution. In all of these cases, gaining the recognition of those involved is a critical part of the problem-solving process.

Evaluating the Solution

Implementation does not necessarily end the problem-solving process. Just as the design process is a circular process with each design leading to possible new designs, problem solving can also be cyclic. Once a solution has been found and implemented, an evaluation should be performed. As with the other steps in the problem-solving process, this step begins with a divergence phase as the problem solver asks what makes a successful solution and how it will be evaluated. To effectively evaluate a solution, a criterion for success must be established. Sometimes there are objective criteria for evaluation but in many circumstances the criterion for success is more subjective.

Once the criterion has been established, the evaluation process should be defined including who will evaluate the solution. Often it is desirable to get a neutral view of someone who was not involved in the formulation of the solution process to be the evaluator.

If the solution is a success, the process may be done. Often in engineering applications, solutions are intermediate and lead to other opportunities. The software industry is a prime example. Software is written to utilize the current computer technologies to address issues. Almost as soon as it is completed, new technologies are available which open up new opportunities or require new solutions.

Sometimes a "success" does not address the true problem as was the case of the turbine disk failures described earlier. In these cases, the evaluation process may need time to deem it a true success. In other cases, solutions have unintended outcomes that require a new solution even if the initial solution was deemed a success.

Whether a solution is effective or not, there is value in taking time to reflect on the solution and its implications. This allows you as a problem solver to learn both from the process and the solution. Critically evaluating solutions and opportunities is a skill that is extremely valuable as an engineer and is discussed further in the critical thinking section of this chapter.

5.5   **Personal Problem-Solving Styles**

The creative problem-solving model presented previously is one of the models that can be used to solve the kind of open-ended problems engineers face in everyday situations. You may find yourself in an organization that uses another model. There are many models for problem solving and each breaks the process up into slightly different steps. Isaken and Treffinger, for example, break the creative problem-solving process into six linear steps [Isaken and Treffinger]. These six steps are:

1. Mess finding
2. Data finding
3. Problem finding
4. Idea finding
5. Solution finding
6. Acceptance finding

Dr. Min Basadur of McMaster University developed another model, a circular one, which he calls Simplex, based on his experience as a product development engineer with the Procter and Gamble Company. This model separates the problem-solving process into eight different steps that are listed below [Basadur].

1. Problem finding
2. Fact finding
3. Problem defining
4. Idea finding
5. Evaluating and selecting
6. Action planning
7. Gaining acceptance
8. Taking action

All these problem-solving processes provide a systematic approach to problem solving. Each process has divergent phases where options need to be generated along with convergent phases when best options need to be selected.

Basadur has created and patented a unique method of helping individuals participating in the creative problem-solving process to identify which parts of the process they are more comfortable in than others. This method, called the Basadur Simplex Creative Problem Solving Profile, reflects your personal creative problem-solving style. Everyone has a different creative problem-solving style. Your particular style reflects your relative preferences for the different parts of the problem-solving process.

Basadur's Creative Problem-Solving Profile method identifies four styles and each style correlates with two of the eight problem-solving steps in the eight step circular model of creative problem-solving above. Thus the eight problem-solving steps are grouped into four quadrants or stages of the complete problem-solving process.

These four steps are shown as the quadrants in Figure 5.4. The Basadur Creative Problem Solving Process begins with quadrant one, the generation of new problems and opportunities. It cycles through quadrant two, the conceptualization of the problem or opportunity and of new, potentially useful ideas and quadrant three, the optimization of new solutions. It ends with quadrant four, the implementation of new solutions. Each quadrant requires different kinds of thinking and problem-solving skills.

Basadur describes the four different quadrants or stages as follows.

Generating

Generating involves getting the problem-solving process rolling. Generative thinking involves gathering information through direct experience, questioning, imagining possibilities, sensing new problems and opportunities, and viewing situations from different perspectives. People and organizations strong in generating skills prefer to come up with options or diverge rather than evaluate and select, or converge. They see relevance in almost everything and think of good and bad sides to almost any fact, idea, or issue. They dislike becoming too organized or delegating the complete problem, but are willing to let others take care of the details. They enjoy ambiguity and are hard to pin down. They delight in juggling many new projects simultaneously. Every solution they explore suggests several new problems to be solved. Thinking in this quadrant includes problem finding and fact finding.

Conceptualizing

Conceptualizing keeps the innovation process going. Like generating, it involves divergence. But rather than gaining understanding by direct experience, it favors gaining understanding by abstract thinking. It results in putting new ideas together, discovering insights that help define problems, and creating theoretical models to explain things. People and organizations strong in conceptualizing skills enjoy taking information scattered all over the map from the generator phase and making sense of it. Conceptualizers need to "understand": to them, a theory must be logically sound and precise. They prefer to proceed only with a clear grasp of a situation and when the problem or main idea is well defined. They dislike having to prioritize, implement, or agonize over poorly understood alternatives. They like to play with ideas and are not overly concerned with moving to action. Thinking in this quadrant includes problem defining and idea finding.

Optimizing

Optimizing moves the innovation process further. Like conceptualizing, it favors gaining understanding by abstract thinking. But rather than diverge, an individual with this thinking style prefers to converge. This results in converting abstract ideas and alternatives into practical solutions and plans. Individuals rely on mentally testing ideas rather than on trying things out. People who favor the optimizing style prefer to create optimal solutions to a few well-defined problems or issues. They prefer to focus on specific problems and sort through large amounts of information to pinpoint "what's wrong" in a given situation. They are usually confident in their ability to make a sound logical evaluation and to select the best option or solution to a problem. They often lack patience with ambiguity and dislike "dreaming" about additional ideas, points of view, or relations among problems. They believe they "know" what the problem is. Thinking in this quadrant includes idea evaluation and selection and action planning.

Implementing

Implementing completes the innovation process. Like optimizing, it favors converging. However, it favors learning by direct experience rather than by abstract thinking. This results in getting things done. Individuals rely on trying things out rather than mentally testing them. People and organizations strong in implementing prefer situations in which they must somehow make things work. They do not need complete understanding in order to proceed and adapt quickly to immediate changing circumstances.

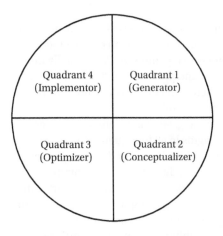

Figure 5.4 The four quadrants of Basadur's Simplex Creative Problem Solving Process (with permission from the Center for Research in Applied Creativity).

When a theory does not appear to fit the facts they will readily discard it. Others perceive them as enthusiastic about getting the job done but also as impatient or even pushy as they try to turn plans and ideas into action. They will try as many different approaches as necessary and follow up or "bird dog" as needed to ensure that the new procedure will stick. Thinking in this quadrant includes gaining acceptance and implementing.

Your Creative Problem-Solving Style

Basadur's research with thousands of engineers, managers, and others has shown that everyone has a different creative problem-solving style. Your particular style reflects your relative preferences for each of the four quadrants of the creative problem-solving process: generating/initiating, conceptualizing, optimizing, and implementing. Your behavior and thinking processes can not be pigeonholed in any single quadrant. Rather they're a combination or blend of quadrants: you prefer one quadrant in particular, but you also have secondary preferences for one or two adjacent quadrants. Your blend of styles is called the creative problem-solving style. Stated another way, your creative problem-solving style shows which particular steps of the process you gravitate toward.

Typically as an engineer, you will be working as part of a team or larger organization. Basadur's research has shown that entire organizations also have their own problem-solving process profiles. An organization's profile reflects such things as the kinds of people it hires, its culture, and its values. For example, if an organization focuses almost entirely on short-term results, it may be overloaded with implementers but have no conceptualizers or generators. The organization will show strengths in processes that deliver its current products and services efficiently. But it will show weaknesses in processes of long-term planning and product development that would help it stay ahead of change. Rushing to solve problems, this organization will continually find itself reworking failed solutions without pausing to conduct adequate fact finding and problem definition. By contrast, an organization with too many generators or conceptualizers and no implementers will continually find good problems to solve and great ideas for products and processes to develop. But it will never carry them to their conclusion. You can likely think of many examples of companies showing this imbalance in innovation process profiles.

Basadur suggests that in order to succeed in creative problem solving, a team requires strengths in all four quadrants. Teams must appreciate the importance of all four quadrants and find ways to fit together their members' styles. Team members must learn to use their differing styles in complementary ways. For example, generating ideas for new products and methods must start somewhere with some individuals scanning the environment, picking up data and cues from customers, and suggesting possible opportunities for changes and improvement. Thus, the generator raises new information and possibilities—usually not fully developed, but in the form of starting points for new projects. Then the conceptualizer pulls together the facts and idea fragments from the generator phase into well-defined problems and challenges and more clearly developed ideas worth further evaluation. Good conceptualizers give sound structure to fledgling ideas and opportunities. The optimizer then takes these well-defined ideas and finds a practical best solution and well-detailed efficient plans for proceeding. Finally, implementers must carry forward the practical solutions and plans to make them fit real-life situations and conditions.

Skills in all four quadrants are equally valuable. As an engineering student, you will find yourself working on group projects. Often, these projects are designs that are open-ended and are well suited for creative problem solving. Use these opportunities to practice utilizing your problem-solving skills and preferences as well as those of your team members.

EXAMPLE 5.6   What is your own preferred problem-solving style? You can find out what your preferred problem-solving style is by taking the Basadur Simplex Creative Problem Solving Profile. This profile and additional information about the Simplex model can be obtained from the Center for Research in Applied Creativity at:

Center for Research in Applied Creativity, 184 Lovers Lane, Ancaster, Ontario, Canada, L9G 1G8 or http://basadursimplex.com.

5.6 **Brainstorming Strategies**

Alex Osborn, an advertising executive in the 1930s, devised the technique of stimulating ideas known as *brainstorming*. Since that time, countless business and engineering solutions have come about as a result of his method. Basically, brainstorming is a technique used to stimulate as many innovative solutions as possible. These then can be used in both analytic and creative problem-solving practices.

The goal of brainstorming is to stimulate your mind to trigger concepts or ideas that normal problem solving might miss. This has a physiological basis. The mind stores information in a network. Memories are accessed in the brain by remembering the address of the memory. When asked about ideas for potential solutions, some would come just by thinking about the problem. A portion of your brain will be stimulated, and ideas will be brought to the surface. However, there will be areas of your brain that will not be searched because the pathways to get there are not stimulated. In brainstorming, we get the brain to search as many regions of the brain as possible.

At times it may seem difficult to generate ideas. Isakssen and Treffinger suggest using the method of idea lists. The concept is simple. Write down the ideas you have currently. Then, use these ideas to branch out into new ideas. To help stimulate your thinking in new directions, a set of key words was developed by Alex Osborn. These words include:

Adapt – What else is like this? What other ideas does this suggest? Are there any ideas from the past that could be copied or adapted? Whom could I emulate?

Put to other uses – Are there new ways to use the object as is? Are there other uses if modified?

Modify – Could you change the meaning, color, motion, sound, odor, taste, form, or shape? What other changes could you make? Can you give it a new twist?

Magnify – What can you add to it? Can you give it a greater frequency? Make it stronger, larger, higher, longer, or thicker? Can you add extra value? Could you add another ingredient? Could you multiply or exaggerate it?

Minify – What could you subtract or eliminate? Could you make it smaller, lighter, or slower? Could you split it up? Could you condense it? Could you reduce the frequency, miniaturize it, or streamline it?

Substitute – What if I used a different ingredient, material, person, process, power source, place, approach, or tone of voice? Who else instead? What else instead? What if you put it in another place or in another time?

Rearrange – Could you interchange the pieces? Could you use an alternate layout, sequence, or pattern? Could you change the pace or schedule? What if you transpose the cause and effect?

Reverse – What if you tried opposite roles? Could you turn it backwards, upside down, or inside out? Could you reverse it or see it through a mirror? What if you transpose the positive and negative?

Combine – Could you use a blend, an assortment, an alloy, or an ensemble? Could you combine purposes, units, ideas, functions, or appeals?

Creativity expert Bob Eberle took Osborn's list and created an easy-to-remember word by adding "eliminate" to the list. Eberle's word is SCAMPER. The words used to form SCAMPER are:

Substitute?
Combine?
Adapt?
Modify? Minify? Magnify?
Put to other uses?
Eliminate?
Reverse? Rearrange?

Using the SCAMPER list may help to generate new ideas.

Individual Brainstorming

Brainstorming can be done either individually or in a group. Brainstorming individually has the advantage of privacy. There may be times when you deal with a professional or personal problem that you don't want to share with a group.

Some research even shows that individuals may generate more ideas by working by themselves. Edward de Bono attributes this to the concept that "individuals on their own can pursue many different directions. There is no need to talk and no need to listen. An individual on his or her own can pursue an idea that seems 'mad' at first and can stay with the idea until it makes sense. This is almost impossible with a group."

When brainstorming individually, select a place that is free from distraction and interruption. Begin by writing down your initial idea. Now you will want to generate as many other ideas as possible. Let your mind wander and write down any ideas that come into your head. Don't evaluate your ideas; just keep writing them down. If you start to lose momentum, refer to the SCAMPER questions and see if these lead to any other ideas.

There are different formats for recording your ideas when brainstorming. Some find generating in ordered lists works best for them. Others will write all over a sheet of paper, or in patterns. Kinesthetic learners often find that putting ideas on small pieces of paper, which allows them to move them around, helps the creative process. Auditory learners might say the ideas out loud as they are generated.

Group Brainstorming

The goal of group brainstorming is the same as with individual brainstorming—to generate as many potential solutions as possible without judging any of them. The power of the group comes when each member of the group is involved in a way that uses their creativity to good advantage.

There are a few advantages to group brainstorming. The first is simply that the additional people will look at a problem differently, and bring fresh perspectives. When brainstorming is done correctly, these different views trigger ideas that you would not have come up with on your own.

Another advantage to group brainstorming is that it gets others involved in the problem-solving process early. As you will see in the following section, a key component in problem solving is implementing your solution. If people whose cooperation you need are in the brainstorming group, they will be much more agreeable to your final solutions. After all, it was partly their idea.

To run an effective group brainstorming session, it helps to use basic guidelines. Those outlined in this section have been proven effective in engineering applications. They are also very helpful for use in student organizations and the committees that engineering students are encouraged to join. This process not only covers the generation of ideas, but also provides a quick way to evaluate and converge on a solution. In some situations, you may want to stop after the ideas are generated and evaluate them later.

The guidelines for this process include:

1. Pick a facilitator
2. Define the problem
3. Select a small group
4. Explain the process
5. Record ideas
6. Involve everyone
7. No evaluating
8. Eliminate duplicates
9. Pick three

Pick a Facilitator

The first step is to select a facilitator who will record the ideas and keep the group focused. The facilitator is also responsible for making sure the group obeys the ground rules of brainstorming.

Define the Problem

It is important that all the participants understand the problem you are looking to solve before generating solutions. Once you start generating ideas, distractions can bring the definition process to a grinding halt. Idea generation would be hampered if the group tried to define the problem they are trying to solve at the wrong time. The definition discussion should happen before the solution idea-generating phase.

Small Group

The group size should be kept manageable. Prof. Goldschmidt from Purdue University recommends groups of six to 12 members. Groups smaller than that can work, but with less benefit from the group process. Groups larger than 12 become unwieldy, and people won't feel as involved. If you need to brainstorm with a larger group, break it into smaller sub-groups and reconvene after the groups have come up with separate ideas.

Explain the Process

Providing the details of the process the group will follow is important. It gives the participants a feeling of comfort knowing what they are getting into. Once the process begins, it is counterproductive to go back and discuss ground rules. This can stymie idea generation just like a mistimed definition of the problem.

Record Ideas

Record ideas in a way that is visible to the whole group. Preferably, arrange the group in a semicircle around the person recording. A chalkboard, flip chart, whiteboard, or newsprint taped to a wall all work well. It is important to record all ideas, even if they seem silly. Often the best ideas come from a silly suggestion that triggers a good idea. By writing the suggestions in a way that is visible to all, the participants can use their sight as well as hearing to absorb ideas which may trigger additional ideas. The use of multiple senses helps to stimulate more ideas. Also, having all the ideas in front of the participants guarantees recollection and allows for new ideas, triggered by earlier ideas. Using the SCAMPER questions is a great way to keep ideas flowing.

Involve Everyone

Start with one idea from the facilitator or another volunteer and write it down. It is easier to get going once the paper or writing surface is no longer completely blank. Go around the group, allowing each person to add one idea per round. If a person doesn't have an idea, let them say "pass," and move on. It is more important to keep moving quickly than to have the participants feel like they must provide an idea every time.

By taking turns, all the participants are ensured an equal opportunity to participate. If the suggestions were taken in a free-flowing way, some participants might monopolize the discussion. The power of group brainstorming lies in taking advantage of the creative minds of every member of the group, not just the ones who could dominate the discussion. Often, the best ideas come from the people who are quiet and would get pushed out of the discussion in a free-flowing setting. It matters not when or how the ideas come, as long as they are allowed to surface.

Continue the process until everyone has "passed" at least once. Even if only one person is still generating ideas, keep everyone involved and keep asking everyone in turn. The one person who is generating ideas might stimulate an idea in someone else at any point. Give everyone the opportunity to come up with a last idea. The last idea may very well be the one that you use.

No Evaluating

This may be hard for the facilitator or some participants. Telling someone that their suggestion is dumb, ridiculous, out-in-left-field, or making any other negative judgment makes that person (and the rest of the group) less likely to speak up. The genius of brainstorming is the free generation of ideas without deciding if they are good until the correct stage. There are countless times when the wackiest ideas are brought up right before the best ideas. Wacky suggestions can trigger ideas that end up being the final solution. Write each idea down, as crazy as some may be. This makes participants feel at ease. They will feel more comfortable making way-out suggestions and will not fear being made fun of or censored.

Avoiding negative comments is one thing, but if you are facilitating, you need to watch for more subtle signals. You might be communicating indirectly to someone that his or her ideas are not good. If you are writing the ideas down, it is natural to say "good idea" or "great idea." What happens, however, if one person gets "great idea" comments and the others don't? Or what if it is just one person who doesn't get the "great idea" comments?

The participants need to feel completely at ease during brainstorming. Providing negative feedback in the form of laughter, sarcastic remarks, or even the absence of praise will work to dampen the comfort level and, therefore, the creativity. It is safest not to make any value judgments about any ideas at this point. If you are facilitating, try to respond to each idea the same way and to keep "order in your court," so to speak. If you are a participant, try not to make value statements or jokes about other ideas.

Eliminate Duplicates

After generating ideas, you may want to study the suggestions. In some cases, it is to your advantage to sort out a solution or the top several solutions. Doing this at the end of group brainstorming can be very easy, and also serves to involve the group members in the decision process.

The first step is to examine the list of ideas and eliminate duplicates. There is a fine line between eliminating duplicates and creating limiting categories. This step is designed to eliminate repeated ideas. Ask if they really are the same. If they are similar, leave both. Creating categories only reduces the possible solutions and defeats the purpose of brainstorming.

The next step is to allow the group to ask clarifying questions about suggestions. This is still not the time for evaluation—only clarification. Don't let this become a discussion of merits. Clarification may help in eliminating and identifying duplicates.

Pick Three

Once everyone understands all ideas as best they can, have them evaluate the suggestions. This can be done quickly and simply. First, ask everyone to pick their three top choices. (Each person can select more if you need more possibilities or solutions.) After each member has determined these in their own minds, let each member vote "out loud" by marking their choices. Do this with a dot or other mark, but not a 1, 2, and 3. The top ideas are selected based on the total number of marked votes they receive. The winners then can be forwarded to the problem-solving phases to see if any survive as the final solution.

Note that with this method of voting, the group's valuable time is optimized. Time is not wasted determining whether an idea deserves a "1" or a "3." Also, no time is spent on the ideas that no one liked. If all ideas are discussed throughout the process, the ideas that are not chosen by anyone or which are least preferred often occupy discussion for quite a while. It's best to eliminate these likely time sinkholes as soon as possible.

Another benefit of not ranking choices at first is that each member could think that their ideas were the fourth on everyone's list, which might frustrate them. Also, how might it affect someone if their ideas were ranked low by the other members? Such a discouraged person could be one who would have very valuable input the next time it is needed, or it could be that this person will be one whom you need as an ally to implement the group's decision. Either way, there is no need to cast any aspersions on ideas which weren't adopted.

5.7   Critical Thinking

Since engineering is so dominated by systematic problem solving, it is easy to get caught up in the methodologies and look at everything as a problem needing a solution or goal. We have presented problem-solving methods that give a systematic approach for analyzing and solving problems. However, many of life's issues are more complex and have no single solution; some cannot be solved. I once assigned a project to design a program using MATLAB, to predict students' financial future and life plans, including a potential spouse. I jokingly suggested that the class use the program when dating to evaluate the financial impact of a potential spouse or life partner. ("Sorry, MATLAB says we wouldn't be compatible. . . .") The class laughed at the suggestion. Clearly, whom you marry, or if you marry, is a complex question and not one that can be solved by some stepwise methodology or computational algorithm. Life is more complex than that.

Issues and challenges you will face have a similar complexity and require a critical perspective to effectively address engineering issues with human, environmental, or ethical impacts. For example, many programs have sought to address poverty, and many programs and "solutions" involved engineering. Poverty, however, has remained a pressing social issue in this country and abroad. Solutions to these larger social issues are complex and often do not have a single or simple solution.

Many of the issues you will be exposed to in your career are problems that cannot be easily or quickly solved. It is easy to get caught up in the processes and methodologies and move to solutions and implementation plans. Sometimes this is appropriate, but other times we need to evaluate our actions and examine the larger issues of our actions. Being able to evaluate why and how you are doing something by taking a step back to gain perspective is an essential skill. At each step in your problem solving, one of the questions you can ask is "why?"

There are times when you need to evaluate your actions, to discover deeper meanings and larger issues. One example involves the design of toys for children who are physically challenged. In the EPICS (Engineering Projects in Community Service, http:// epics.ecn .purdue.edu) Program, students work on engineering solutions to local community issues. One of the partnerships in the program is with a clinic that works with children with disabilities. As the students work with the children, families, and therapists, they gain an appreciation for the myriad complex issues facing these children. When the engineering students design toys, they must design them so they can be used with the given physical abilities of the children. This is fairly straightforward.

What the students have learned is that they can address many other issues, such as socialization dimensions, in their designs. Some of the toys are designed so that children with disabilities can more easily interact with their peers in their integrated classroom, when using the toys. The students have also learned to design so that the children with disabilities can control aspects of the play when using the toys with their peers. The engineering students have learned about the socialization of children with and without disabilities, and its impact on self-esteem. They have also come to understand the role of play in child development and the barriers children with disabilities may encounter. Armed with this knowledge, the engineering students can design toys that meet a wider spectrum of needs.

When working with issues such as children or adults with disabilities, there are not simple problems to be solved, but complex issues that need to be addressed and pondered. By working through these issues and their multiple dimensions, students can become aware of the social and emotional issues of their work. They can become better designers, and better educated about relevant issues.

As engineers, we have grave responsibilities because our work has great potential for good and for harm. While having the highest of ethical standards is essential, keeping a critical eye on the context and direction of your work is equally vital. Creating opportunities to gain perspective and evaluating your actions, experiences, and perceptions is as important as the problem-solving methodology presented in this text. At the beginning of the creative problem-solving process, as well as the design process, we discussed how crucial it was to get the question right. Similarly, when evaluating an issue, asking the right question will enable you to generate the best solution.

We use our personal background and experience to evaluate problems, and this perspective can limit our thinking. When evaluating a solution, getting an outside opinion has value. You must become aware of the perspective you bring to issues based on your background so it does not limit you. Thomas Edison wanted people around him who would not limit their thinking by jumping to conclusions. He would take every prospective engineer to lunch at his favorite restaurant, which was known for their soup, which he would recommend. Invariably, the candidate would order soup. Edison wanted to see if he or she would salt the soup before tasting it. Salting without tasting, he reasoned, was an indication of someone who jumps to conclusions. He wanted people who were open to questioning. Periodically, ask yourself if there are things that you are salting without tasting. Are there issues for which you jump to conclusions based on your background, personality, or thinking style?

Critical thinkers ask the Why? question about themselves as well as about the things around them. Periodically take a step back and ask Why? or What? about yourself. Why are you studying to become an engineer? What will you do as an engineer to find personal fulfillment? Becoming a critical thinker will make you a better problem solver and engineer, and one who is more engaged in society and more fulfilled.

REFERENCES

Basadur, M., *Simplex: A Flight to Creativity*, Creative Education Foundation, Inc., 1994.

Beakley, G. C., Leach, H. W., Hedrick, J. K., *Engineering: An Introduction to a Creative Profession*, 3rd Edition, Macmillan Publishing Co., New York, 1977.

Burghardt, M.D., *Introduction to Engineering Design and Problem Solving*, McGraw-Hill Co., 1999.

Eide, A.R., Jenison, R. D., Mashaw, L. H., Northup, L. L., *Introduction to Engineering Problem Solving*, McGraw-Hill Co., 1998.

Fogler, H. S. and LeBlanc, S. E., *Strategies for Creative Problem Solving*, Prentice Hall, Inc., Englewood Cliffs, New Jersey, 1995.

Gibney, Kate, *Awakening Creativity*, Prism, March 1998, pp. 18-23.

Isaksen, Scott G. and Treffinger, Donald J., *Creative Problem Solving: The Basic Course*, Bearly Limited, Buffalo, New York, 1985.

Lumsdaine, Edward and Lumsdaine, Monika, *Creative Problem Solving—Thinking Skills for a Changing World*, 2nd Edition, McGraw-Hill Co, 1990.

Osborn, A. F., *Applied Imagination: Principles and Procedures of Creative Problem-Solving*, Charles Scriber's Sons, New York, 1963.

Panitx, Beth, B*rain Storms*, Prism, March 1998, pp. 24-29.

EXERCISES AND ACTIVITIES

5.1   A new school has exactly 1000 lockers and exactly 1000 students. On the first day of school, the students meet outside the building and agree on the following plan: the first student will enter the school and open all the lockers. The second student will then enter the school and close every locker with an even number (2, 4, 6, 8, etc.). The third student will then "reverse" every third locker (3, 6, 9, 12, etc.). That is if the locker is closed, he or she will open it; if it is open, he or she will close it. The fourth student will then reverse every fourth locker, and so on until all 1000 students in turn have entered the building and reversed the proper lockers. Which lockers will finally remain open?

5.2   You are stalled in a long line of snarled traffic that hasn't moved at all in twenty minutes. You're idly drumming your fingers on the steering wheel, when you accidentally started tapping the horn. Several sharp blasts escaped before you realize it. The driver of the pickup truck in front of you opens his door, gets out and starts to walk menacingly toward your car. He looks big, mean, and unhappy. Your car is a convertible and the top is down. What do you do?

5.3   Your school's team has reached the national championship. Tickets are very difficult to get but you and some friends have managed to get some. You all travel a great distance to the site of the game. The excitement builds on the day of the game until you discover that your tickets are still back in your room on your desk. What do you do? What is the problem that you need to solve?

5.4   You are a manufacturing engineer working for an aircraft company. The production lines you are responsible for are running at full capacity to keep up with the high demand. One day, you are called into a meeting where you learn that one of your suppliers of bolts has been forging testing results. The tests were never done, so no data exists to tell if the bolts in question meet your standards. You do not know how long the forging of the test results has been going on. You are only told that it has been "a while." This means that your whole production line including the planes ready to be delivered may be affected. You are asked to develop a plan to manage this crisis. What do you do?

Some background information:

- Bolts are tested as batches or lots. A specified number are taken out of every batch and tested. The testing typically involves loading the bolts until they fail.
- The bolts do not have serial numbers so it is impossible to identify the lot number from which they came.
- The supplier who forged the tests supplies 30% of your inventory of all sizes of bolts.
- Your entire inventory is stored in bins by size. Each bin has a mixture of manufacturers. (To reduce dependence on any one company, you buy bolts from three companies and mix the bolts in the bins sorted by size.)
- Bolts have their manufacturer's symbol stamped on the head of the bolt.
- It takes weeks to assemble or disassemble an aircraft.
- Stopping your assembly line completely to disassemble all aircraft could put your company in financial risk.
- Your customers are waiting on new planes and any delays could cause orders to be lost.
- The FBI has arrested those who forged the test and their business has been closed, at least temporarily.

5.5   How much of an automobile's tire wears off in one tire rotation?

5.6   A farmer going on a trip with a squirrel, acorns, and a fox had to cross a river in a
        boat in which he couldn't take more that one of them with him each time he crossed.
        Since he often had to leave two of them together on one side of the river or the other,
        how could he plan the crossings so that nothing gets eaten, and they all get across the
        river safely?

5.7   Measure the height of your class building using two different methods. Compare your
        answers and indicate which is more accurate.

5.8   Pick a grassy area near your class building. Estimate how many blades of grass are in
        that area. Discuss your methodology.

5.9   How high do the letters on an expressway sign need to be to be readable?

5.10   A company has contacted you to design a new kind of amusement park ride. Generate
        ideas for the new ride. Which idea would you recommend?

5.11   Estimate the speed of a horse.

5.12   Estimate the maximum speed of a dog.

5.13   Assume that an explosion has occurred in your building. Your room is intact but
        sustained damage and is in immediate danger of collapsing on you and your
        classmates. All exits but one (as specified by your instructor) are blocked. The one
        remaining is nearly blocked. Quickly develop a plan to get your classmates out. Who
        goes first and who goes last? Why?

5.14   A child's pool is eight feet in diameter and two feet high. It is filled by a garden hose up
        to a level of one foot. The children complain that it is too cold. Can you heat it up to an
        acceptable temperature using hot water from the house? How much hot water would
        you need to add?

5.15   How else could you heat the water for the children in problem 5.14?

5.16   Without referring to a table or book, estimate the melting temperature of aluminum.
        What are the bounds of potential melting temperatures? Why? What is the actual
        melting temperature?

5.17   Select a problem from your calculus class and use the analytic problem-solving steps
        to solve it.

5.18   Select a problem from your chemistry class and use the analytic problem-solving steps
        to solve it.

5.19 Select a problem from your physics class and use the analytic problem-solving steps to solve it.

5.20 Write a one to two-page paper on a historical figure that had to solve a difficult problem. Provide a description of the problem, solution, and the person.

5.21 Aliens have landed on earth and they come to your class first. They are very frustrated because they can not communicate with you. Develop a plan to break this communication barrier and show them that we are friendly people.

5.22 You are working on a project that has had many difficulties that are not your fault. However, you are the project leader and your manager has become increasingly frustrated with you and even made a comment that you are the kind of irresponsible person who locks your keys in your car. Your manager is coming into town to see first hand why the project is having problems and you are going to pick her up. You leave early to take time to relax. With the extra time, you decide to take the scenic route and stop the car next to a bubbling stream. Leaving your car, you go over to the stream to just sit and listen. The quiet of the surroundings is just what you needed to calm down before leaving for the airport. Glancing at your watch, you realize that it is time to leave to pick up your manager. Unfortunately, you discover that you indeed have locked your keys in the car. A car passes on the road about every 15 minutes. The nearest house is probably a mile from your location and the nearest town is 15 miles away. The airport is 20 miles away and the flight is due to arrive in 30 minutes. What will you do?

5.23 You are sitting at your desk in the morning reading your e-mail when your boss bursts into your office. It seems that a local farmer was not happy with your company and has dropped a truckload of potatoes in your plant's entrance. The potatoes are blocking the entrance and must be moved. You have been selected as the lucky engineer to fix this problem. What would you do to move the potatoes and how would you get rid of them? Brainstorm ideas and select the best solution. (Note: You work at a manufacturing plant of some kind that does not use potatoes in its process. The potatoes also appear to be perfectly fine, just in the way.)

5.24 You work in a rural area that is known for chickens. An epidemic has swept through the area killing thousands of chickens. The EPA has mandated that the chickens may not be put into the landfill nor can they be buried. Develop a plan to get rid of the chickens.

5.25 A fire has destroyed thousands of acres of forest. Seedlings have been planted in the area but are being destroyed by the local wildlife. Develop a way to reforest the region by preventing the wildlife from eating the seedlings.

5.26 A survey shows that none of the current high school students in your state are planning to major in engineering when they get to college. Your dean has asked your class to fix this dilemma (with no students, she would be out of a job!). Develop a plan to convince the high school students that engineering is worthwhile.

5.27   Select a problem from your math book. How many ways can you solve the problem? Demonstrate each method.

5.28   You and seven of your friends have ordered a round pizza. You need to cut it into eight pieces and are only allowed to make three straight cuts. How can you do this?

5.29   Select an innovative product and research the story of its inventors. Prepare a presentation on how the ideas were formed and the concept generated.

5.30   With which part of the Basadur problem-solving process are you most comfortable? Write a one-page paper on how your preference affects your problem solving.

5.31   Take the Basadur problem-solving profile test. Report on your scores. How will this information affect your ability to solve problems?

5.32   Organizations have problem-solving styles. Select an organization and prepare a brief report on their own problem-solving styles.

5.33   You and your classmates have discovered that your funding for college will be withdrawn after this semester. Develop ideas on how you could obtain the money for the rest of your college expenses.

5.34   Your class has been hired by a new automotive company that wants to produce a brand new car. Their first entry into the market must be truly innovative. Do a preliminary design of this car. What features would it have to set it apart from current vehicles?

5.35   An alumnus of your engineering program has donated money to build a jogging track that would encircle your campus. The track will be made of a recycled rubberized material. How much of this rubberized material will be needed?

5.36   Every Christmas season it seems that one toy becomes a "I have to have one" craze. You and some classmates have been hired to come up with next year's toy of the season. What is it?

5.37   A professional sports team has decided to locate in your town. A local businessperson is going to build a stadium for the team. Unfortunately, the site is contaminated. The project will fall through unless you can devise a way to get rid of the contaminated soil under the new arena. What do you do with the soil?

5.38   NASA has decided to send astronauts to the other planets in the solar system. Psychologists, however, have determined that the astronauts will need something to keep them occupied during the months of travel or they will develop severe mental problems that would threaten the mission. Your job is to devise ways to keep the astronauts occupied for the long journey. Remember that you are limited to what will fit in a space capsule.

5.39   Landfill space is rapidly running out. Develop a plan to eliminate your city's dependence on the local landfill. The city population is 100,000.

5.40   An appliance manufacturer has hired you to expand their market. Your job is to develop a new household appliance. What is it and how will it work?

5.41   College students often have trouble waking up for early classes. Develop a system that will guarantee that the students wake up and attend classes.

5.42   Select one of your classes. How could that class be more effective in helping you learn?

5.43   Prepare a one-page report on a significant engineering solution developed in the past five years. Evaluate its effectiveness and report on the outcomes (both intended and unintended) of this solution.

5.44   Habitat for Humanity International spearheaded an initiative to eliminate substandard housing within the Georgia county in which it is headquartered. One of the first steps was to develop a set of standards to determine if housing was substandard or not. Develop criteria for your own community for determining whether a housing unit is substandard.

5.45   If your own community were to undertake an initiative like the one presented in the previous question, what would need to be done (both from an engineering and community perspective)?

5.46   Identify one current topic related to engineering or technology that has no single simple solution and write a brief paper discussing issues related to the topic.

5.47   Brainstorm ideas for toys for physically disabled children. Narrow your choices to the top three and identify the benefits of each design. What are the important issues that you would need to consider in your designs?

5.48   In the critical thinking section in this chapter, engineers had to pass Edison's "salt test." Identify areas in your own thinking where you might tend to "salt before tasting" (where you might jump to conclusions or use your own biases to overlook potential options).

5.49   Critically evaluate why you are majoring in engineering. What do you hope to gain by studying engineering?

5.50   Write a one-page paper on your responsibilities to society as a citizen in your community and as an engineer.

5.51   Early one morning is starts to snow at a constant rate. Later, at 6:00 a.m., a snow plow sets out to clear a straight street. The plow can remove a fixed volume of snow per unit time. In other words, its speed it inversely proportional to the depth of the snow. If the plow covered twice as much distance in the first hour as the second hour, what time did it start snowing?

5.52   While three *wise* men are asleep under a tree a mischievous boy paints their foreheads red. Later they all wake up at the same time and all three start laughing. After several minutes suddenly one stops. Why did he stop?

5.53   Using just a 5-gallon bucket and a 3-gallon bucket, can you put four gallons of water in the 5-gallon bucket? (Assume that you have an unlimited supply of water and that there are no measurement markings of any kind on the buckets.)

5.54   A bartender has a three-pint glass and a five-pint glass. A customer walks in and orders four pints of beer. Without a measuring cup but with an unlimited supply of beer how does he get a single pint in either glass?

5.55   You are traveling down a path and come to a fork in the road. A sign lays fallen at the fork indicating that one path leads to a village where everyone tells the truth and the other to a village where everyone tells lies. The sign has been knocked down so you do not know which path leads to which village. Then someone from one of the villages (you don't know which one) comes down the path from which you came. You may ask him one question to determine which path goes to which village. What question do you ask?

5.56   Four mathematicians have the following conversation:

Alice: I am insane.
Bob: I am pure.
Charlie: I am applied.
Dorothy: I am sane.
Alice: Charlie is pure.
Bob: Dorothy is insane.
Charlie: Bob is applied.
Dorothy: Charlie is sane.

You are also given that:

- Pure mathematicians tell the truth about their beliefs.
- Applied mathematicians lie about their beliefs.
- Sane mathematicians' beliefs are correct.
- Insane mathematicians' beliefs are incorrect.

Describe the four mathematicians.

5.57   Of three men, one man always tells the truth, one always tells lies, and one answers yes or no randomly. Each man knows which man is which. You may ask three yes/no questions to determine who is who. If you ask the same question to more than one person you must count it as a question used for each person asked. What three questions should you ask?

5.58 Tom is from the census bureau and greets Mary at her door. They have the following conversation:

Tom: I need to know how old your three kids are.
Mary: The product of their ages is 36.
Tom: I still don't know their ages.
Mary: The sum of their ages is the same as my house number.
Tom: I still don't know their ages.
Mary: The younger two are twins.
Tom: Now I know their ages! Thanks!

How old are Mary's kids and what is Mary's house number?

5.59 Reverse the direction of the fish image below by moving only three sticks:

Computer Tools for Engineers

6.1 The Internet

Introduction

Assume that you are a food process engineer, and a dairy company has hired you for a project:

"Design a process to manufacture high-quality powdered milk for a profit."

The first step required to solve the dairy design problem is to fill gaps in your knowledge base. That is, you need to do research so that you have a better understanding of the design problem. Some things you should learn are:

- How is powdered milk manufactured?
- What are the consumer safety and health issues involved?
- Who manufactures equipment for dairy processing?
- What other dairy companies manufacture powdered milk?
- Who does dairy processing research?

Research, under the best conditions, can take a significant amount of time. So your aim is to avoid excessive trips to the library and the use of bound indexes. It is preferable to access as much information as possible via your computer. The Internet provides such capabilities.

History of the Internet

The Internet is a worldwide network of computers that allows electronic communication with each other by sending data packets back and forth. The Internet has been under development since the 1960s. At that time, the Department of Defense was interested in testing out a concept that would connect government computers. It was hoped that such a network would improve military and defense capabilities (Glass and Ables, 1999). Before the end of 1969, machines at the University of Southern California,

Table 6.1 Selection of Top-Level Domain Names

Name	Category
edu	U.S. Educational
gov	U.S. Government
COM	U.S. Commercial
mil	U.S. Military
org	Non-profit Organization
net	Network
XX (e.g., fr, uk)	Two-Letter Country Code (e.g., France, United Kingdom)

Stanford Research Institute, the University of California at Santa Barbara, and the University of Utah were connected.

The linking of those four original sites generated much interest in the Internet. However, extensive customization was required to get those original sites linked, because each site used its own operating system. It was apparent that the addition of more sites would require a set of standard protocols that would allow data sharing between different computers. And so the 1970s saw the birth of such protocols.

Protocols are used to move data from one location to another across the network, and verify that the data transfer was successful. Data are sent along the Internet in *packets*. Each packet may take a different route to get to its final destination. The *Internet Protocol* (IP) guarantees that if a packet arrives at its destination, the packet is identical to the one originally sent. A common companion protocol to the IP is the *Transmission Control Protocol* (TCP). A sending TCP and a receiving TCP ensure that all the packets that make up a data transfer arrive at the receiving location.

Internet applications were also developed in the 1970s and are still used today. The Telnet program connects one computer to another through the network. *Telnet* allows a user on a local computer to remotely log into and use a second computer at another location. The *ftp* (file transfer protocol) program allows users to transfer files on the network. Both applications were originally non-graphical; they require that text commands be entered at a prompt.

Internet IP addresses were established in the 1970s. All organizations that become part of the Internet are assigned a unique 32-bit address by the Network Information Center. IP addresses are a series of four eight-bit numbers. For instance, IP addresses at Purdue University take the form 128.46.XXX.YYY, where XXX and YYY are numbers between zero and 255.

In the 1980s, corporations began connecting to the Internet. With so many sites joining, a new naming system was needed. In came the *Domain Naming Service*. This hierarchical naming system allowed easier naming of individual hosts (servers) at a particular site. Each organization connected to the Internet belongs to one of the top-level domains (Table 6.1). Purdue University falls under the U.S. educational category; therefore, it has the domain name purdue.edu. Once an organization has both an IP address and a domain name, it can assign addresses to individual hosts. This author's host name is pasture.ecn.purdue.edu. "Pasture" is a large computer (server) on the Purdue University Engineering Computer Network (ecn), a network of engineering department servers.

As *Internet Service Providers* (ISP), such as America On-Line and CompuServe, came into being in the 1990s, the Internet became more mainstream. The development of the browser named Mosaic made accessing information on the Internet easier for the general public. A *browser* is software that allows you to access and navigate the Internet. Mosaic displays information (data) using both text and graphics. Each page of information is written using the *HyperText Markup Language* (HTML). The most useful aspect of HTML is the hyperlink. A *hyperlink* connects two pages of information. The beauty of the hyperlink is that it allows an Internet user to move through information in a non-linear fashion. That is, the user can "surf" the *World Wide Web* (WWW). The Web, not to be confused with the Internet, refers to the interconnection of HTML pages. Two of the most common browsers are Netscape Navigator and Microsoft Internet Explorer. Updates to these browsers have allowed for the construction of more complicated web pages, such as multiple-frame pages. Sound and video can now be played through the use of helper applications. Greater user interaction with the Web has been made possible through HTML-coded references to Java applets (defined later).

Searching the Web

Let's return to the powdered-milk problem. You'd like to use the Internet to answer some of your research questions. First focus on searching for equipment vendors. Assume that you already know about one company that supplies food processing equipment, called APV. Check to see what kinds of equipment they have available for dairy processing. Using your Web browser, you access APV's website by providing the URL (*Universal Resource Locator*) for APV's home page. In other words, you tell the Web browser what Web address to use to access APV's information. Assume that the URL is "http://www.apv.com." This URL consists of the two primary components of a Web address: the protocol (means for transferring information) and the Internet address. The protocol being used here is http (Hypertext Transport Protocol) which allows access to HTML. The Internet address consists of the host name (apv) and the domain (com).

APV's URL takes you to APV's introductory (home) page. Once there, you can browse their website for information that pertains to your specific research problem. At one point, you find yourself on a page that describes a one-stage spray dryer designed specifically for temperature-sensitive materials such as milk. A spray dryer is a piece of equipment that converts liquid into a powder, or as in your application, milk to powdered milk. The URL for this page is http://www.apv.com/anhydro/spray/onestage.htm. This Web address is longer than that for APV's home page. The address contains a path name (/anhydro/spray/) and a file name (onestage.htm). The path name tells you the location within APV's directory structure of the file that you are accessing with your Web browser, just as you store files in directories on your hard drive.

APV is just one equipment vendor. You'd like to investigate others, but you do not know any other Web addresses. To further your research, you use a Web search engine. A *search engine* matches user-specified keywords with entries in the search engine's database. The engine then provides a list of links to the search results. There are a number of search engines such as Yahoo! (www.yahoo.com) and Google (www.google.com). A keyword search performed using one search engine will yield different results than the same keyword search on another search. The reason for the different results lies in the type of information that a particular search engine tends to have in its database.

You decide to use the search engine Google to do a keyword search for "spray dryer." When your search results come up, you immediately notice that your query was not specific enough. You get back links to sites on laundry dryers! There are so many sites on the Internet that well-constructed queries are a must. Each search engine provides links to searching tips. In your case, you find that a keyword search for "spray dryer and food" improves your search results.

Home Pages

Home pages are Web pages developed by users that contain information about topics of interest to the user, and often include links to the user's favorite websites. The language of all Web pages is HTML. An *HTML document* is a text file containing information the user wants posted to the Web, and instructions on how that information should look when the Web page is accessed by a browser. HTML code can be written from scratch using a text editor, or a composer can be used to generate the code. Many word processors and desktop publishers now have the capability of saving files in HTML format so that the document is displayed by the Web browser exactly as it was in the word-processing program. The HTML code consists of instructional tags as seen in Figure 6.1. This code generates the Web page shown in Figure 6.2.

Your employer may ask you to design a Web page discussing the process you developed through the completion of the powdered-milk process design project. A personal reason for creating your own home page is to make available your resume and samples of your work to all Web users. This sort of visibility may result in employment opportunities such as internships and post graduation positions.

More information on creating home pages can be found on the Web or in any book on HTML or the Internet. You may find "HTML: The Complete Reference" by T. A. Powell (1998) to be helpful when writing HTML code.

Libraries & Databases on the Internet

Not all information is available through the Internet but an ever-increasing amount is. New search features are coming on-line each year and databases are continually emerging and growing. The Internet can be used to locate many sources for research through search features such as Google Scholar (scholar.google.com). Information that is not available through the web can be located using the Internet such as through a website of a university's library. Most have sophisticated search tools that can be used to access articles in electronic form or can locate hard copies.

Contacting People through the Internet via E-mail

The Internet can also be used to send electronic mail (e-mail) to individuals, groups, or mailing lists. E-mail is currently exchanged via the Internet using the *Simple Mail Transport Protocol* (SMTP) and *Multipurpose Internet Mail Extension* (MIME). To send e-mail, two things are required: an e-mail package and e-mail addresses. There are many e-mail packages from which to choose.

Regardless of the brand, e-mail tools have certain common functions including an ability to send mail, receive mail, reply to mail, forward mail, save mail to a file, and delete mail.

To communicate via the Internet, an *e-mail address* is required. All e-mail users are assigned a unique e-mail address that takes the form username@domain.name.

```
<HTML>
<HEAD>
<TITLE>FOODS BLOCK LIBRARY – BASIC ABSTRACT</TITLE>
</HEAD>

<BODY BGCOLOR="#FFFFFF">
<H1>A Computer-Aided Food Process Design Tool for
Industry, Research, & Education</H1>
<H2>Heidi A. Diefes, Ph.D.</H2>
<H2><I>Presented at the Food Science Graduate Seminar<BR>
Cornell University, Fall 1998</I></H2>
<HR ALIGN=LEFT NOSHADE SIZE=5 WIDTH=50%>
<P>
There is a great need in the food industry to link food science research with food
engineering and process technologies in such a way as to facilitate an increase in
product development efficiency and product quality. Chemical engineers achieve this
through industry standard computer-aided flowsheeting and design packages. In
contrast, existing programs for food processing applications are limited in their abil-
ity to handle the wide variety of processes common to the food industry. This
research entails the development of a generalized flowsheeting and design program
for steady-state food processes which utilizes the design strategies employed by food
engineers.
</P>
</BODY>
</HTML>
```

Figure 6.1 HTML code for Web page shown in Figure 6.2.

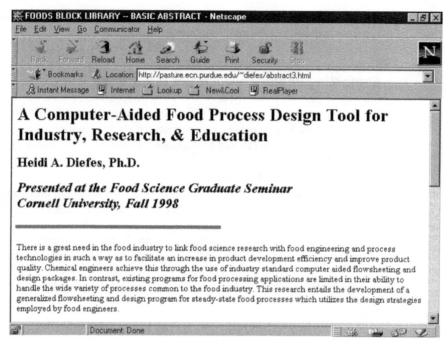

Figure 6.2 Web page generated using HTML code in Figure 6.1.

However, it can be difficult to locate a person's e-mail address. One method of locating an e-mail address is to use one of the many people-finding websites. For example, the Netscape People Finder can be used to find both home and e-mail addresses. In addition, most colleges and universities have methods for searching for staff and students on the school's website.

Once you have selected an e-mail address and e-mail software, and after you have signed up with an ISP, you can send and receive e-mail. As part of your dairy design problem, you may keep in touch with the dairy company contacts and other engineering consultants via e-mail. A typical e-mail message might read like the example that follows.

To: dairyconsult@purdue.edu
cc: dairycontacts@purdue.edu
Attachments: spdry.bmp
Subject: Spray dryer design model
Message:
I have been working with the spray dryer vendor on a model of the system we intend to use. Attached is a current picture of the model. Please advise.
H. A. Diefes

The above e-mail message contains both the typical required and optional components. An e-mail is sent:

To: username@domain.name or an alias (explained in the next paragraph). A copy
(cc:) may also be sent to a username@domain.name or an alias.
Attachments: are files to be appended to the main e-mail message.
Subject: is a brief description of the message contents.
Message: is the actual text of the message.

An *alias* is a nickname for an individual or group of individuals, typically selected by the "owner." An e-mail to an alias gets directed to the actual address. Address books, another common feature of e-mail packages, allow the user to assign and store addresses and aliases. For example, in your address book you may group all the e-mail addresses for your dairy company contacts under the alias "dairy contacts." If you put "dairy contacts" in the To or cc line of your mail tool, the e-mail message will be sent to everyone on the "dairy contacts" list.

Words of Warning about the Internet

With millions of people having access to the Internet, one must be very discerning about the information one accesses. Common sense and intuition are needed to determine the validity of material found on the Web.

In addition, care must be taken not to plagiarize materials found on the Web. As with hard copy materials, always cite your sources.

Further, computer viruses can be spread by executable files (.exe) that you have copied. The virus is unleashed when you run the program. Text files, e-mail, and other non-executable files cannot spread such viruses. Guard against viruses attached to executable files by downloading executable files only from sites that you trust. Anti-virus programs such as those from Norton or McAfee can be installed to help detect and protect against viruses.

There are clues that you can use to ascertain the reliability of the website. One is the type of the URL domain name. Pages that have the extensions such as .edu, indicating an educational institution, or .gov, indicating some kind of governmental agency, should have more credibility than a random .com or .net site. Be aware that just because an extension such as .edu is in the URL, does not mean that the institution sponsors that webpage. A "~" symbol is used to designate a particular user is in control of the page rather than the institution. The example in Figure 6.2 showed a Web page that has the URL pasture.ecn.purdue .edu\~diefes\abstract3.html. The "~diefes" means that this is a page under the control of the user "diefes" and not the host institution.

Other clues to the trustworthiness of a website include the information on the Web page. Is there an author cited on the page? Is that author credible? What is the creation date of the site? Are there references on the site? The Internet has made information readily available but it has also made it easy to post information that is not credible, which places a responsibility on you as the researcher to verify information as credible.

Social media, such as Facebook and MySpace, and other websites such as YouTube have rapidly expanded how information is transferred by text, pictures and video. Many students use these tools to share information and update others on what they are doing or have done. We recommend caution when using the Internet to share information. Once released, you lose control over that content. Once something is on the Internet, it may be there for a very long time. Employers, scholarship committees, and graduate schools can and will do Internet searches on candidates and there are many stories of students who posted something they thought was funny to their friends but ended up being viewed as evidence of their professionalism, or lack thereof. The Internet is a powerful tool that has fundamentally changed the way we live and communicate but like any tool it needs to be used with caution.

6.2 Word-Processing Programs

Throughout the powdered-milk process design project, you will need to do various types of writing. The largest document will be the technical report for the design. You also may have to complete a non-technical report for management. In addition, you may write memorandums and official letters, and you will need to document experimental results. The computer tool that will enable you to write all these papers is a *word processor*, such as Microsoft Word.

The most basic function of a word processor is to create and edit text. When word processors first became available in the early 1980s, that was about the extent of their ability. The user could enter, delete, copy, and move text. Text style was limited to bold, italic, and underline, and text formats were few in number.

"Word processing," as understood today, is really a misnomer, since the capabilities of word processors have become indistinguishable from what was once considered desktop publishing. Word processors such as Microsoft Word, Corel WordPerfect, and Adobe FrameMaker have features that are particularly useful to engineers. Equation, drawing, and table editors are features which make the mechanics of technical writing relatively easy. Most programs include spelling and grammar checkers and an on-line thesaurus to help eliminate common writing errors.

Figure 6.3 shows the graphical user interface (GUI) for Microsoft Word. All word processors display essentially the same components. The title bar indicates the application name and the name of the file that is currently open. The menu bar provides access to approximately 10 categories of commands such as file, edit, and help commands. Some word processors have a toolbar that gives direct access to commands that are frequently used such as "save," "bold," and "italic." A typical word processor GUI also has scroll bars and a status bar. The scroll bars allow the user to move quickly through the document. The status bar provides information on current location and mode of operation. The workspace is where the document is actually typed. The workspace in Figure 6.3 shows a snippet of the technical information related to the spray dryer of our example.

This textbook you are reading was written using word processors. Throughout these pages you can find examples of tables, equations, and figures created using a word

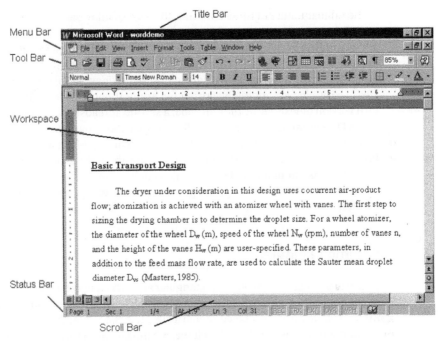

Figure 6.3 Example of a Microsoft Word paper in progress.

processor. A wide variety of typefaces, paragraph styles, and formats are also in evidence. You also will see images and pictures that have been imported into this document. A user may import materials that are created using other computer applications as well.

Documenting is an ongoing process. You will need to return time and again to your word processor to record all subsequent steps of the dryer design process. Periodically, you will use the word processor to write memos and official letters and to prepare experimental reports, technical documents, and non-technical papers.

6.3 Spreadsheets

Introduction

Much data collection is performed during the research and design phases of an engineering project. Data come from a variety of sources including experiments, design calculations, product specifications, and product statistics. While designing the powdered-milk process, you have gathered cost information from your equipment vendors and collected experimental data on powder solubility. You may like to use the cost information in an economic analysis of your process (recall that your process is supposed to make a profit). You may also attempt to analyze your laboratory data. Due to the tabular nature of these data sets, this sort of information is best recorded with a spreadsheet.

Spreadsheet packages have their origins in finance. The idea for the first spreadsheet application, called VisiCalc, came in 1978 from a Harvard Business School student who was frustrated by the repetitive nature of solving financial-planning problems (Etter, 1995). VisiCalc was originally written for Apple II computers. IBM came out with its own version of VisiCalc in 1981. In 1983, a second spreadsheet package was released by Lotus Development Corporation, called Lotus 1-2-3. Now there are a number of other spreadsheet packages from which to choose, including Microsoft Excel, Corel Quattro Pro for PC, and Xess for UNIX. One thing to note about the evolution of the spreadsheet is the increase in functionality. VisiCalc centered on accounting math: adding, subtracting, multiplying, dividing, and percentages. Spreadsheets today have many, many different functions.

Common Features

A spreadsheet or worksheet is a rectangular grid which may be composed of thousands of columns and rows. Typically, columns are labeled alphabetically (A, B, C, . . . , AA, AB, . . .) while rows are labeled numerically (1, 2, 3, . . .). At the intersection of a given column and row is a cell. Each cell has a column-row address. For example, the cell at the intersection of the first column and first row has the address A1. Each cell can

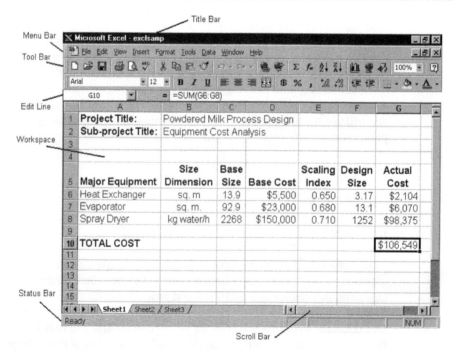

Figure 6.4 Example of a Microsoft Excel spreadsheet.

behave like a word processor or calculator in that it can contain formattable text, numbers, formulas, and macros (programs). Cells can perform simple tasks independently of each other, or they can reference each other and work as a group to perform more difficult tasks.

Figure 6.4 shows the GUI for Microsoft Excel. Notice that many of the interface features are similar to those of a word processor, such as the title bar, menu bar, toolbar, scroll bars, and status bar. Many menu selections are similar as well, particularly the "File" and "Edit" menus. The standard features of all spreadsheets are the edit line and workspace. The edit line includes the name box and formula bar. The name box contains the active cell's address or user-specified name. Within the worksheet in Figure 6.4 is the beginning of the equipment costing information and economic analysis of our example.

Today's spreadsheets have many features in common including formatting features, editing features, built-in functions, data manipulation capabilities, and graphing features. Let's look at each one of these areas individually.

As with word processing, the user has control over the format of the entire worksheet. Text written in a cell can be formatted. The user has control over the text's font, style, color, and justification. The user can also manipulate the cell width and height as well as border style and background color.

Editing features are also similar to a word processor. Individual cell contents can be cleared or edited. The contents of one or more cells can be cleared, copied, or moved. Excel even has a spell checker.

Spreadsheets have categories of built-in functions. At a minimum, spreadsheets have arithmetic functions, trig functions, and logic functions. Arithmetic functions include a function to sum the values in a set (range) of cells (SUM) and a function to take an average of the values in a range of cells (AVERAGE). Trig functions include SIN, COS, and TAN, as well as inverse and hyperbolic trig functions. Logic (Boolean) functions are ones that return either a true (1) or false (0) based on prescribed conditions. These functions include IF, AND, OR, and NOT. You might use a logic function to help you mark all equipment with a cost greater than $100,000 for further design consideration. The formula to check the heat exchanger cost would be IF(G6>100000, "re-design", "okay"). This formula would be placed in cell H6. If the value in cell G6 is greater than 100,000, the word *re-design* is printed in cell H6. Otherwise, the word *okay* is placed in cell H6.

Spreadsheet developers continually add built-in, special purpose functions. For instance, financial functions are a part of most spreadsheets. An example of a financial function is one that computes the future value of an investment. In addition, most spreadsheets have a selection of built-in statistical functions.

When a user wants to perform repeatedly a task for which there is not a built-in function, the user can create a new function called a macro. A *macro* is essentially a series of statements that perform a desired task. Each macro is assigned a unique name that can be called whenever the user has need of it.

Spreadsheets also provide a sort function to arrange tabular data. For instance, vendor names can be sorted alphabetically, while prices can be sorted in order of increasing or decreasing numeric value.

Graphing is another common spreadsheet feature. Typically, the user can select from a number of graph or chart styles. Suppose that you are interested in studying ways to reduce the energy it takes to operate the spray dryer. Figure 6.5 is a Microsoft Excel-generated *x–y* plot of such an analysis. Here you are investigating the impact that recycling hot air back into the dryer has on steam demand.

Advanced Spreadsheet Features

Database construction, in a limited fashion, is possible in some spreadsheet applications such as Excel. A database is a collection of data that is stored in a structured manner that makes retrieval and manipulation of information easy. A database consists of records, each having a specified number of data fields. Typically, the user constructs a data entry form to facilitate entry of records. Once a database has been constructed, the user can sort and extract information. You could use the database feature to catalog all expenses during implementation of your powdered-milk process. This would require a record for each piece of equipment purchased. Data fields would include size, quantity, list price, discounts, and actual cost of each item.

Real-time data collection is another advanced feature of spreadsheets. This feature allows data and commands to be the received from, and sent to, other applications or computers. During the design phase of the powdered-milk process, you might use this feature to log data during pilot plant experiments. Later, you might use this feature to create a spreadsheet that organizes and displays data coming from on-line sensors in your powdered-milk process. This would enable line workers to detect processing problems quickly.

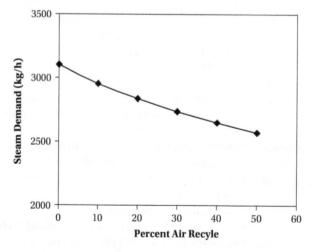

Figure 6.5 An x-y plot generated using Microsoft Excel's Chart Wizard.

6.4 **Mathematics Software**

Mathematical modeling is a large component of the design process. For the pow-dered-milk process, you may need to develop a mathematical model of parts of the process. Problems of this nature are best solved using mathematical or computational software.

MATLAB

MATLAB is a scientific and technical computing environment. It was developed in the 1970s by the University of New Mexico and Stanford University to be used in courses on matrix theory, linear algebra, and numerical analysis (MathWorks, 1995). While ma-trix math is the basis for MATLAB ("MATrix LABoratory"), the functionality of MATLAB extends much farther. MATLAB is an excellent tool for technical problem solving. It contains built-in functions and a command window, used to perform computational operations. (Taken from "Introduction to Technical Problem Solving with MATLAB" by Great Lakes Press, www.glpbooks.com.) Simply put, MATLAB can be used like a calcu-lator; it can be used like a programming language; it can be used as a data analysis tool.

We recommend an excellent text for those interested in learning to use MATLAB: "Introduction to Technical Problem Solving with MATLAB" by Jon Sticklen.

A typical MATLAB session would resemble that in Figure 6.6. The plot generated by these commands is shown in Figure 6.7. At this point, don't concern yourself with the details, but just get a feel for the use of MATLAB.

Note that MATLAB has a text-based interface. On each line, one task is completed before going on to the next task. For instance, on the first line, the value of Q is set. On the second line, the variable R is set equal to a range of values from 4×10^{-6} to 7×10^{-6} in steps of 1×10^{-7}. On the third line, the arithmetic operation $Q \times R$ is performed. On command lines 4 through 6, MATLAB's built-in plotting commands are used to gener-ate a plot of the results.

The commands shown in Figure 6.6 could be saved as a file with a ".m" extension (MATLAB file) and run again and again, or edited at a future date. MATLAB has a

```
>       Q = 1.5e6;
>       R = 4e-6:1e-7:7e-6;
>       A = Q.*R;
>       plot(R,A,'s-');
>       xlabel('Area m^2');
>       ylabel('Resistance m^2/W');
```

Figure 6.6 MATLAB code for determining heat transfer surface area.

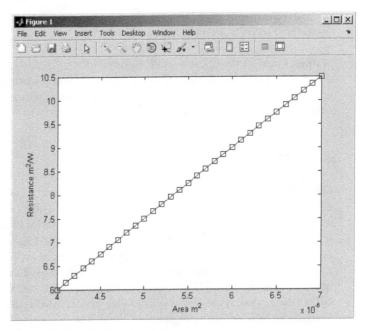

Figure 6.7 Sample MATLAB plot.

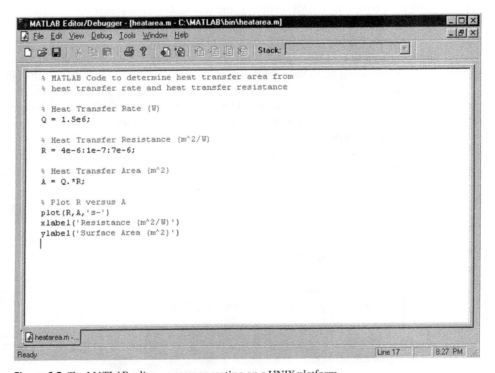

Figure 6.8 The MATLAB editor as seen operating on a UNIX platform.

built-in editor/debugger interface that resembles a very simple word processor (Figure 6.8). The debugger helps the user by detecting syntax errors that would prevent the proper execution of the code. The editor shown in Figure 6.8 is displaying the code for the heat transfer area calculation.

MATLAB maintains, and periodically adds, toolboxes, which are groups of specialized built-in functions. For example, one toolbox contains functions that enable image and signal processing. There is also a Symbolic Math Toolbox that enables symbolic computations based on Maple (see next paragraph). SIMULINK is another toolbox, a graphical environment for simulation and dynamic modeling. The user can assemble a personal toolbox using MATLAB script and function files that the user creates.

MathCad and Maple

MathCad and Maple are graphical user interface-based mathematics packages. Each is 100% WYSIWYG—computer jargon for "What You See Is What You Get." That is, what the user sees on the screen is what will be printed. There is no language to learn since mathematical equations are symbolically typed into the workspace. The workspace is unrestrictive—text, mathematics, and graphs can be placed throughout a document. These packages are similar to a spreadsheet in that updates are made immediately after an edit occurs. MathCad and Maple feature built-in math functions, built-in operators, symbolic capabilities, and graphing capabilities. Perhaps the greatest advantage of MathCad is its ability to handle units. In MATLAB, the user is charged with keeping track of the units

Figure 6.9 Example of a MathCad window and problem-solving session.

associated with variables. In MathCad, units are carried as part of the assigned value. Figure 6.9 shows the GUI for MathCad and the text used to solve the problem.

6.5   **Presentation Software**

At regular intervals during the design process, presentations must be made to research groups and to management. Presentation programs allow a user to communicate ideas and facts through visuals. Presentation programs provide layout and template designs that help the user create slides.

PowerPoint

Microsoft PowerPoint is a popular electronic presentation package. Figure 6.10 shows a slide view window of PowerPoint. As you have seen in other Microsoft programs, the GUI for this package has a title bar, a menu bar, a toolbar, and a status bar. The unique feature of this tool is the drawing toolbar.

PowerPoint has five different viewing windows. The "slide view" window displays one slide at a time. Typically, this is the window through which slide editing is done.

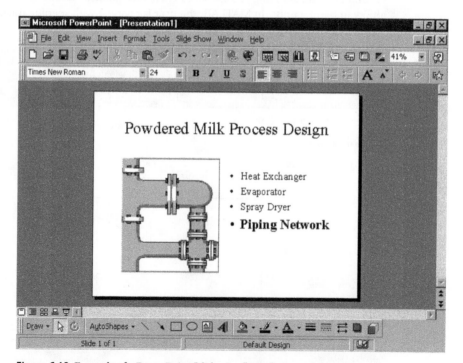

Figure 6.10 Example of a PowerPoint "slide view" window.

The "outline view" window lists the text for each slide in a presentation. The "slide sorter view" window graphically displays all the slides. As the name implies, slide sorting is done through this window. The "notes page view" allows the user to attach text with each slide. Since bullets are used on most slides, this feature comes in very handy. It allows the user to come back to a presentation file at a future date and still remember what the presentation covered. The slides can be viewed sequentially using the "slide show view." This is the view used during a presentation.

Presentation packages have a number of features that help the user put together very professional presentations. PowerPoint has a gallery of templates. Templates specify color scheme and arrangement of elements (e.g., graphics and text) on the slide. Figure 6.10 shows a "slide view" with Clip Art. PowerPoint also has animation capabilities. With the proper computer hardware, sound and video clips also can be incorporated into a presentation.

6.6 **Operating Systems**

An *operating system* is a group of programs that manage the operation of a computer. It is the interface between the computer hardware, the software, and the user. Printing, memory management, and file access are examples of processes that operating systems control. The most recognized operating systems are UNIX, originally from AT&T Bell Laboratories; MS-DOS, from Microsoft Corporation; Microsoft's Windows family (95, 98, NT, etc.); and the Apple OS for Macintosh computers.

EXERCISES AND ACTIVITIES

6.1   Access your school's home page; surf the site and locate information about your department.

6.2   Perform a keyword search on your engineering major using two different Web search engines.

6.3   Determine if your school has on-line library services. If so, explore the service options.

6.4   Send an e-mail to a fellow student and your mom or dad.

6.5   Search for the e-mail address of an engineering professional society.

6.6   Find the website of an engineering society. Prepare a one-page report on that organization based on the information found at their website.

6.7   Select the website of the publisher of this text (www.oup.com). Prepare a critique of the website itself. What could be done to improve it? How does it compare with other websites you have accessed? If you wish, you may send your suggestions to the publisher via e-mail.

6.8   Prepare an *oral* presentation on the reliability or unreliability of information obtained from the Internet. Include techniques to give yourself confidence in the information.

6.9   Prepare a *written* report on the reliability or unreliability of information obtained from the Internet. Include techniques to give yourself confidence in the information.

6.10   Do a case study on a situation where inaccurate information from the Internet caused a problem.

6.11   Write a paper for one of your classes (e.g., Chemistry) using the following word processor features: text and paragraph styles and formats.

6.12   Create a table in a word processor.

6.13   Prepare a resume using a word processor.

6.14   Prepare two customized resumes using a word processor for two different companies you might be interested in working for someday.

6.15   Prepare a two-page report on why you selected engineering, what discipline you wish to enter, the possible career paths you might follow, and why. Create the first page using a word processor. Create the second page using a different word processor.

6.16   Prepare a brief presentation on your successes or frustrations using the two different word processors. Which would you recommend?

6.17   Use a spreadsheet to record your expenses for one month.

6.18   Most spreadsheets have financial functions built in. Use these to answer the following question: If you now started saving $200 per month until you were 65, how much money would you have saved? Assume an annual interest rate of: a) 4%, b) 6%, and c) 8%.

6.19   Assume that you wish to accumulate the same amount of money on your 65th birthday as you had in the previous problem, but you don't start saving until your 40th birthday. How much would you need to save each month?

6.20   A car you wish to purchase costs $25,000; assume that you make a 10% down payment. Prepare a graph of monthly payments versus interest rate for a 48-month loan. Use a) 6%, b) 8%, c) 10%, and d) 12%.

6.21   Repeat the previous problem, but include plots for 36- and 60-month loans.

6.22   Prepare a brief report on the software tools used in an engineering discipline.

6.23   Interview a practicing engineer and prepare an *oral* presentation on the computer tools he or she uses to solve problems.

6.24   Interview a practicing engineer and prepare a *written* report on the computer tools he or she uses to solve problems on the job.

6.25   Interview a senior or graduate student in engineering. Prepare an *oral* presentation on the computer tools he or she uses to solve problems.

6.26   Interview a senior or graduate student in engineering. Prepare a *written* report on the computer tools he or she uses to solve problems.

6.27   Select an engineering problem and prepare a brief report on which computer tools could be used in solving the problem. Which are the optimum tools, and why? What other tools could be used?

6.28   Prepare a brief report on the kinds of situations for which engineers find the need to write original programs to solve problems in an engineering discipline. Include the languages most commonly used.

6.29   Prepare a presentation using PowerPoint which discusses an engineering major.

6.30   Prepare a presentation which speculates on what computer tools you will have available to you in five, 10, and 20 years.

CHAPTER 7

Teamwork

7.1 Introduction

More than ever, engineering schools are requiring their students to work in teams. These teams take the form of collaborative study groups, laboratory groups, design groups as part of individual classes, and design groups participating in extracurricular competitions such as the solar car, human powered vehicle, SAE formula car, concrete canoe, bridge design, Rube Goldberg projects, and many others. This team emphasis in engineering schools mirrors a management philosophy of the use of teams that has swept through the corporate world in the past few decades. Engineering employers, in fact, are the principal drivers for the use of teams in engineering schools. They seek engineering graduates who have team skills as well as technical skills. The purpose of this chapter is to impart a team vision, to show why teams and collaboration are so important to organizational and technical success, and to give practical advice for how to organize and function as a team. If you study this chapter, work through the exercises, and view some of the references, you should have most of the tools you need to be a successful team member and leader while in engineering school. Apply the principles in this chapter to your team experiences while in college and learn from and document each team experience. By so doing, you will have a portfolio of team experiences that will be a valuable part of your resume. Above all, you will have the satisfaction that comes from being a part of great teams. Teams are fun.

7.2 Engineers Often Work in Teams

A full-page advertisement in *USA Today*, October 14, 1991, states clearly the current highly competitive climate in industry and what Chrysler (now DaimlerChrysler), for example, is doing about it. The use of interdisciplinary teams is an important part of the solution:

> No more piece-by-piece, step-by-step production. Now it's teams. Teams of product and manufacturing engineers, designers, planners, financing and marketing people—together from the start. . . . It's how we built Dodge Viper . . . from dream

to showroom in three years, a record for U.S. car makers. From now on, all our cars and trucks will be higher quality, built at lower cost and delivered to the market faster. That's what competition is all about.

The greater the problem, the greater the need for the use of teams. Individuals acting alone can solve simple problems, but tough problems require teams. This section lists and then discusses the reasons many organizations have embraced the use of teams. Why is the use of teams so popular today?

- Greater understanding of the power of collaboration.
- Engineers are asked to solve extremely complex problems.
- More factors must be considered in design than ever before.
- Many corporations are international in scope, with design and manufacturing engineering operations spread across the globe.
- Because time to market is extremely important to competitive advantage, "concurrent engineering," inherently a team activity, is widely employed.
- Corporations are increasingly using project management principles.

Power of Collaboration

There is a greater understanding today than ever before of the importance and power of creative collaboration. One recent book on the subject is *Organizing Genius: The Secrets of Creative Collaboration*, by Warren Bennis and Patricia Ward Biederman. Their first chapter is titled "The End of the Great Man," in which they argue that great groups are able to accomplish more than talented individuals acting alone, stating that "None of us is as smart as all of us." They cite a survey by Korn-Ferry (the world's largest executive search firm) and *The Economist* in which respondents were asked who will have the most influence on global organizations in the next ten years. Sixty-one percent of the respondents answered "teams of leaders," whereas 14 percent said "one leader." Individual chapters of their book focus on "great groups." A chapter on the Manhattan Project discusses the development of the atomic bomb during the Second World War by a team of great scientists. One chapter discusses Lockheed's "Skunk Works," developers of the U-2 spy plane, the SR-71 Blackbird, and the F117-A Stealth fighter-bomber used in the Persian Gulf war. Another example is the development of the graphical user interface at Xerox's PARC (Palo Alto Research Center) and its adoption into the Macintosh computer by Steve Jobs and Steve Wozniak. Steve Jobs' goal was to lead his team to create something not just great, but "insanely great," to "put a dent in the universe."

Another recent book on collaboration is *Shared Minds: The New Technologies of Collaboration*, by Michael Schrage. He defines collaboration as "the process of *shared creation:* two or more individuals with complementary skills interacting to create a shared understanding that none had previously possessed or could have come to on their own." He cites examples of collaboration in science (Watson and

Crick; Heisenberg, Bohr, Fermi, Pauli, and Schrodinger; Einstein and mathematician Marcel Grossman), music (Gilbert and Sullivan; Rodgers and Hart; McCartney and Lennon), art (Monet and Renoir; Van Gogh and Gauguin; Picasso and Braque), and literature (editor Maxwell Perkins and writers F. Scott Fitzgerald and Thomas Wolfe; Ezra Pound and T. S. Eliot).

James Watson and Stanley Crick, winners of the Nobel prize for discovering the double helix structure of DNA credit their collaboration for their success: "Both of us admit we couldn't have done it without the other—we were interested in what the answer is rather than doing it ourselves." Crick wrote that "our . . . advantage was that we had evolved . . . fruitful methods of collaboration." Sir Isaac Newton, father of classical physics, in a letter to his contemporary (and rival) Robert Hooke, recognizing his debt to fellow scientists, wrote: "If I have seen further (than you and Descartes) it is by standing on the shoulders of Giants." Modern physics also found its genesis in collaboration. Heisenberg, Bohr, Fermi, Pauli, and Schrodinger, all Nobel laureates for their roles in founding quantum physics, worked, played, and vacationed together as they developed together their revolutionary new ideas.

Today's Complex Problems

Engineers are asked to solve increasingly complex problems. The complexity of mechanical devices has grown rapidly over the past two hundred years. David Ullman, in his text *The Mechanical Design Process*, gives the following examples. In the early 1800s, a musket had 51 parts. The Civil War–era Springfield rifle had 140 parts. The bicycle, first developed in the late 1800s, has over two hundred parts. An automobile has tens of thousands of parts. The Boeing 747 aircraft, with over 5 million components, required over ten thousand person-years of design time. Thousands of designers worked over a three-year period on the project. Most modern design problems involve not only many individual parts, but also many subsystems—mechanical, electrical, controls, thermal, and others—each requiring specialists acting in teams.

Many Design Factors

Engineering designers must consider more factors than ever before, including initial price, life cycle costs, performance, aesthetics, overall quality, ergonomics, reliability, maintainability, manufacturability, environmental factors, safety, liability, and acceptance in world markets. Satisfying these criteria requires the collective teamwork of design and manufacturing engineers, marketing, procurement, business, and other personnel. The typical new engineering student, being interested primarily in technical things, tends to believe that engineers work isolated from other factors, always seeking the "best technical solution." This is not true. The very act of engineering involves solving difficult problems, finding technical solutions while considering numerous constraints. Engineers make things happen, they are doers, they solve problems, while interacting with many entities within the culture of complex organizations. Often acting at organizational boundaries, engineers are technical experts who, to be truly effective, must understand economics, management, marketing, and the dynamics of the business enterprise.

International Scope

Many corporations are international in scope, with design and manufacturing engineering operations spread across the globe. These operations require teams to work together who may never physically meet. Instead, they meet and share data via electronic means. For example, it is not unusual for design engineers in the United States to collaborate on a design with Japanese engineers for a product that will be manufactured in China. Manufacturing personnel from the Chinese plant also contribute to the project. As another example, software engineers in the U.S., the U.K., and India may collaborate on a software project literally around the clock and around the world. At any time during a 24-hour period, programmers in some part of the world will be working on the product.

Time to Market and Concurrent Engineering

"Concurrent engineering," inherently a team activity, is widely employed in order to achieve better designs and to bring products to market more quickly. Time to market is the total time needed to plan, prototype, procure materials, create marketing strategies, devise tooling, put into production, and bring to market a new product. In the traditional company, each of these steps is done serially, one at a time. It is a relay race, with only one person running at a time, passing the baton to the next runner. The design engineer "tosses a design over the wall" to a manufacturing engineer who may not have seen it yet. She then "tosses it over the wall" to procurement personnel, and so on. Concurrent engineering, on the other hand, is a parallel operation. Like rugby, all of the players run together, side-by-side, tossing the ball back-and-forth. Marketing, manufacturing, and procurement personnel are involved from the beginning of the design phase. The use of teams, as well as new technologies such as CAD/CAM, rapid prototyping, shared data, and advanced communications have radically changed the process of engineering. By the use of teams and CAD/CAM technologies, in the face of withering international competition, U.S auto makers have cut the time needed to bring a new car to market from approximately 5 years to less than 3 years.

Speed—that is, timely delivery of products to the marketplace—is critical for companies to be profitable. Jack Welch, General Electric's widely respected CEO, has said, "we have seen what wins in our marketplaces around the globe: speed, speed, and more speed." Xilinx Corporation has stated that their research shows that a six-month delay in getting to market reduces a product's profitability by a third over its life cycle. An October 25, 1991 *Business Week* article stated: "Reduce product

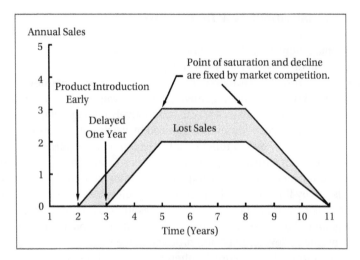

Figure 7.1 The cost of delay in bringing an equivalent new product to market. (Used by permission of Charles J. Nuese.)

development time to one third, and you will triple profits and triple growth." Figure 7.1, shown below, shows graphically the losses incurred over the lifetime of a product due to delays in bringing the product to market. Early product introduction leads to higher profits because early products bring higher profit margins, are more apt to meet customer needs, and are more likely to capture a larger market share. [This figure was created by Dr. Charles J. Nuese and included in his wisdom-laden book, *Building the Right Things Right*, on product development.] The figure is fictitious, but it is based on principles Dr. Nuese has observed during his career as an engineer in and as a consultant to the semiconductor industry.

Developing new products is a key to growth and profits. Rubbermaid Corporation, during a recent year, introduced 400 new products—more than one per calendar day! At least one third of Rubbermaid's $2 billion in annual sales come from products created in the previous 5 years. 3M Corporation CEO L. D. DiSimone, facing flat revenues and stiff competition, adopted a business focus of generating 30 percent of revenues from products less than 4 years old. The creative thought processes, the innovation needed to create a steady stream of new ideas, and the expertise to manufacture and bring these to market quickly, are all functions that teams perform well.

It is important for engineering students to understand the importance of speed (without compromising quality). Delays have many negative consequences. Engineering professors who supervise teams should provide incentives to students for early completion of team projects, or at least penalties for failure to meet deadlines. Student teams should recognize the importance of speed, push themselves and their teams to early completion of projects, and document these on their resumes and/or portfolios.

Project Management is a Team Approach

Project management is widely practiced in industries and government labs. Many graduate engineers discover on their first job that engineers frequently work in project-oriented teams. Unfortunately, most engineering students never take a project management class. Project management principles were developed in the defense industry in the 1950s and 1960s as a way to productively manage Department of Defense contracts. The typical corporation is organized by vertical divisions or lines, where individuals are clustered by job functions. For example, a corporation may have separate research, manufacturing, engineering, human resources, product development, marketing, and procurement divisions. Project management is a way of organizing individuals not by function but by products or projects. Selected individuals from research, engineering, manufacturing, human resources, procurement, and other divisions are gathered together as a team for a particular purpose, for example solving a corporate problem, developing a new product, or meeting a crucial deadline. A project management approach, therefore, is inherently a cross-functional team approach and an excellent way to solve problems. Successful project management, however, is not an easy task.

The project manager's job is to complete a project, on time, within budget, with the personnel that he is given. Unfortunately, he is rarely able to hand pick his personnel, and his supervisors and/or customers typically set the project completion time and budget. In other words, project managers are virtually never given all the people, time, and money they wish for. These tough constraints are often mirrored in student design teams. Their teacher picks the team members, they don't have much money or resources to work with, and their teacher determines the time frame. These conditions are uncomfortable for students, but they prepare students for the engineering world.

Project managers use tools that any team leader would benefit from learning how to use. Project managers plan work requirements and schedules and direct the use of resources (people, money, materials, equipment). One popular project management tool that student teams should learn to use is the Gantt chart (a simple Gantt chart is pictured in Table 7.1). A Gantt chart organizes and presents clearly information on task division and sequencing. Tasks are divided into sub-tasks, the shading conveys the sequence in which the tasks should be completed and how long a task is expected to take, and this version shows person(s) responsible for each task.

Project managers monitor projects by tracking and comparing progress to predicted outcomes. They use project management software such as Microsoft Project for planning and monitoring purposes. Project managers carefully guard the scope of the project from excessive creep. A project is successful if it is completed on time, within budget, at the desired quality, with effective utilization of resources. For more information on project management, the reader is encouraged to consult any text on Project Management and to view the publication titled "PMBOK: Project Management Body of Knowledge" online at www.pmi.org, the web site of the Project Management Institute.

Table 7.1 Team A-9 ("Semi-Conscious Objectors"): Gantt Chart for Object Relocator: Fall 2013

Task	Leader(s)	Oct 11–17	Oct 18–25	Oct 26–31	Nov 1–7	Nov 8–14	Nov 15–21	Nov 22–28	Nov 29 –Dec 5	Dec 6–12
Finalize Design	Team	▩								
Procure Materials	Joe		▩							
Pick-Up Module	Jaime			▩						
Conveyor Assembly	George				▩					
Working Prototype	Team					▩				
Fine Tuning Machine	Jaime/Team					▩	▩			
Instruction Manual	George						▩			
Professor Tests Projects	Team				▩	▩				
Report Rough Draft	Sally							▩		
Finalize CAD Drawings	Ahmad							▩		
Presentation	Ahmad/Joe								▩	
Final Report Due	Team									▩
Judges test best two projects per class	Team									▩

Nov

7.3   **Team Growth Stages**

Simply forming teams to solve problems may seem like a good idea, given the evidence and cases from the previous section, but teams are not a cure-all. Teams require nurturing. Successful team development is a process. Whether in a corporate, academic, or other setting, all teams must pass through several developmental phases before they truly become productive. These stages are titled *forming, storming, norming, performing, and adjourning.* Unfortunately, successfully completing the first four stages and becoming a productive, performing team is not guaranteed. *Many teams in practice never reach the performing stage;* they remain mired in the forming, storming, or norming stages. Every team's challenge is to grow through these stages and achieve "performance."

Stage 1: Forming

The first stage of team development is "forming." In this stage, typical dialogue may consist of the following: "Nice to meet you. Yeah, I'm not sure either why we are here. I'm afraid this might be a lot of work." In the formation stage, team members get acquainted with one another, with the leader (or choose a leader if one is not pre-appointed), with the team's purpose, and with the overall level of commitment (work load) required. Team members begin to learn of one another's personalities, abilities, and talents, but also of one another's weaknesses and idiosyncrasies. Individual team members are typically shy, reserved, self-conscious, and uncertain. A key role of the leader in this stage is to lead team ice-breaker activities, facilitate discussion, encouraging all to speak and quieting some who might tend to dominate the conversation. Another role of the leader is to help the team begin to focus on the task at hand.

Stage 2: Storming

The second stage of team development is "storming." In this stage, typical dialogue (or private thoughts) may consist of the following: "Do I have to work with this team? What did I do to deserve this? There clearly aren't any super-heroes on this team, including that dizzy leader. How are we supposed to solve this messy problem?" During the storming stage, the enormity and complexity of the task begins to sink in, sobering and discouraging the participants. "We are supposed to do what? By when?" Teams are rarely formed to solve easy problems, only very difficult and complex ones. Typically, time schedules are unrealistically short, and budgets are inadequate. Further complicating the issue, teammates have learned enough about their fellow team members to know that there are no super-heroes, no saviors they can count on to do it all. (One person doing all the work is a team failure.) Some team members may not initially "hit it off" well with the others. Cliques or factions may emerge within the team, pitted against others. Since the leader's weaknesses (all leaders have weaknesses) are by now apparent, some individuals or factions may vie for leadership of the team. Though possibly under siege, the leader's role is critical during the storming stage. The leader must help the team to focus on its collective strengths, not weaknesses, and to direct their energies toward the task. To be a successful team, it is not necessary for team members to like one another or to be friends. A professional knows how to work productively with individuals with widely differing backgrounds and personalities. Everyone must learn the art of constructive dialogue and compromise.

Stage 3: Norming

The third stage of team development is "norming." In this stage, typical dialogue (or private thoughts) may consist of the following: "You know, I think we can do it. True, there are no super-heroes, not by a long shot, but once we stopped fighting and started listening to one another, we discovered that these folks have some good ideas. Now if we can just pull these together . . ." "Norms" are shared expectations or rules of conduct. All groups have some kinds of norms, though many times unstated. Do you recall a time you joined a group or team and felt subtle influence to act, dress, look, speak, or work in a particular way? The more a team works together, they tend to converge to some common perspectives and behaviors. During the norming stage, team members begin to accept one another instead of complaining and competing. Rather than focusing on weaknesses and personality differences, they acknowledge and utilize one another's strengths. Individual team members find their place in the group and do their part. Instead of directing energies toward fighting itself, the team directs its collective energy toward the task. The key to this shift of focus is a collective decision to behave in a professional way, to agree upon and adhere to norms. Possible norms include working cooperatively as a team rather than individually, agreeing on the level of effort expected of everyone, conducting effective discussions and meetings, making effective team decisions, and learning to criticize one another's ideas without damaging the person. One suggested norm is that all team members are expected to be at all meetings, or to communicate clearly in the event that he/she cannot attend. During the norming stage, feelings of closeness, interdependence, unity, and cooperation develop among the team. The primary role of the leader during the norming phase is to facilitate the cohesion process. Some team members will lag behind the team core in embracing norms. The leader and others on the team (at this stage leadership is beginning to emerge from others on the team as well) must artfully nudge individuals along toward group accountability and focus on the task.

Stage 4: Performing

The fourth stage of team development is "performing." In this stage, typical dialogue (or private thoughts) may consist of the following: "This is a fun team. We still have a long way to go, but we have a great plan. Everyone is pulling together and working hard.

No super-heroes, but we're a super team." In this stage, teams accomplish much. They have a shared, clear vision. Responsibilities are distributed. Individual team members accept and execute their specific tasks in accordance to the planned schedule. They are individually committed, and hold one another accountable. On the other hand, there is also a blurring of roles. Team members "pitch in" to help one another, doing whatever it takes for the team to be successful. In a performing team, so many team members have taken such significant responsibility for the team's success that the spotlight is rarely on a single leader anymore. Typically, whoever initially led the team is almost indistinguishable from the rest of the team.

Stage 5: Adjourning

The fifth and final stage of team development is "adjourning." Because teams are typically assembled for a specific purpose or project, there is a definite time when the team is disbanded when that goal is accomplished. If the team was successful, there is a definite feeling of accomplishment, even euphoria, by the team members. On the other hand, an under-performing team will typically feel anger or disappointment upon adjournment.

These stages, originally set forth by Bruce W. Tuckman in 1965, appear in some form in most books which discuss teams. Some leave off the final step, adjourning. Others switch the order and give the steps as forming, conforming (instead of "norming"), storming, and performing.

7.4   **What Makes a Successful Team?**

Have you ever been a member of a great team? It's a great experience, isn't it? On the other hand, how many of you have ever been part of an unsuccessful team? It was a disappointing experience, wasn't it? How do you measure or evaluate a team's success or lack thereof? What specific factors make a team successful? What factors cause a team to fail?

First it is important to define what a team is and is not. A team is not the same as a group. The term "group" implies little more than several individuals in some proximity to one another. The term "team," on the other hand, implies much more. A team is two or more persons who work together to achieve a common purpose. The two main elements of this definition are "purpose" and "working together."

All teams have a purpose and a personality. A team's purpose is its task, the reason the team was formed. Its personality is its people and how the team members work together. Different teams have their own style, approach, dynamic, and ways of communicating that are different than other teams. For example, some teams are serious, formal, and businesslike, whereas others are more informal, casual, and fun-loving. Some teams are composed of friends, others are not. Friendship, though desirable, is not a necessary requirement for a team to be successful. Commitment to a common purpose and to working together is required. Regardless of style or personality, a team must have professionalism. A team's professionalism ensures that its personality promotes productive progress toward its purpose.

Attributes of a Successful Team

The successful team should have the following attributes:

1. A common goal or purpose. Team members are individually committed to that purpose.
2. Leadership. Though one member may be appointed or voted as the team leader, ideally every team member should contribute to the leadership of the team.
3. Each member makes unique contributions to the team's project. A climate exists in the team that recognizes and appropriately utilizes the individual talents/abilities of the team members.
4. Effective team communication: Regular, effective meetings; honest, open discussion. Ability to make decisions.
5. Creative spark. There's excitement and creative energy. Team members inspire, energize, and bring out the best in one another. A "can do" attitude. This creative spark fuels collaborative efforts and enables a team to rise above the sum of its individual members.
6. Harmonious relationships among team members. Team members are respectful, encouraging, and positive about one another and the work. If conflicts arise, there are peacemakers on the team. The team's work is both productive and fun.
7. Effective planning and use of resources. This involves appropriate breakdown of the tasks and effective utilization of resources (people, time, money).

Team Member Attributes

Individual team members of the successful team should have these attributes:

1. Attendance: Attends all team meetings, arriving on time or early. Dependable, like clockwork. Faithful, reliable. Communicates in advance if cannot attend a meeting.
2. Responsible: Accepts responsibility for tasks and completes them on time, needing no reminders nor cajoling. Has a spirit of excellence, yet is not overly perfectionistic.
3. Abilities: Possesses abilities the team needs, and contributes these abilities fully to the team's purpose. Does not withhold self or draw back. Actively communicates at team meetings.
4. Creative, Energetic. Acts as an energy source, not a sink. Brings energy to the team. Conveys a sense of excitement about being part of the team. Has a "can do" attitude about the team's task. Has creative energy and helps spark the creative efforts of everyone else.

5. Personality: Contributes positively to the team environment and personality. Has positive attitudes, encourages others. Acts as a peacemaker if conflict arises. Helps the team reach consensus and make good decisions. Helps create a team environment that is both productive and fun. Brings out the best in the other team members.

7.5 Team Leadership

The Need for Leadership

Every successful human endeavor involving collective action requires leadership. Great teams need great leadership. Without leadership, humans tend to drift apart, act alone, and lose purpose. They may work on the same project, but their efforts, without synchronization and coordination, interfere with rather than build on one another. Without coordinated direction, people become discouraged, frustrations build, and conflicts ensue. Money is wasted, time schedules deteriorate, but the greatest loss is a loss of human potential. A failed team effort leaves a bitter taste not easily forgotten. Incomparably sweet, on the other hand, is the thrill of team success.

The Leader's Role

The leader ensures that the team remains focused on its purpose and that it develops and maintains a positive team personality. He challenges and leads the team to high performance and professionalism. With one hand he builds the team, with the other he builds the project. In support of these objectives, the leader must do the following:

1. Focus on the purpose: Help the team remain focused on its purpose.
2. Be a team builder: The leader may actively work on some project tasks himself, but his most important task is the team, not the project. He builds, equips, and coordinates the team so they can accomplish their purpose and succeed as a team.
3. Plan well and effectively utilize resources (people, time, money). He assesses and utilizes effectively the team members' abilities.
4. Run effective meetings. Ensure that the team meets together regularly and that the meetings are productive.
5. Communicate effectively. The leader effectively communicates the team vision and purpose. He also effectively praises good work and gives effective guidance for improvement of substandard performance.
6. Promote team harmony by fostering a positive environment. If team members focus on one another's strengths instead of weaknesses, conflicts are less likely to arise. Conflict is not necessarily evil, however. The effective leader must not be afraid of conflict. He should view conflicts as an opportunity to improve team performance and personality, and to refocus it on its purpose.

7. Foster high levels of performance, creativity, and professionalism. Have a high vision. Challenge team members to do the impossible, to think creatively, and to stimulate one another to high performance.

The Entire Team Leads, Too

Every team needs all the leadership it can get. Most teams have a single individual who is either voted or appointed as its leader. In the best teams, however, many members contribute leadership in ways that support and complement the appointed leader. Like the appointed leader, every member should focus on the purpose, build the team, plan, recognize the gifts of others, contribute to effective meetings, communicate well, promote a harmonious team environment, and be creative. Ideally, every member, together with the leader, should build the team with one hand and labor on the project with the other hand. Noted author John Maxwell defines leadership in this way: "Leadership is influence." Every team member, from his/her position on the team, has the opportunity to considerably influence the team's performance. How important is the leadership from the entire team? It is possible to have a truly outstanding leader, but if the team chooses to use its influence to not follow the leader, to undermine the leader, and to promote team disunity, the team will fail. Conversely, a team with a relatively weak appointed leader can be greatly successful if the team members pitch in and use their influence to build the team. Ultimately, the entire team—both the members and the appointed leader—must work together to build a great team. It takes a team to make a great team.

Leadership Styles

According to Schermerhorn, "leadership style is a recurring pattern of behaviors exhibited by a leader." It is the tendency of a leader to act or to relate to people in a particular way. There are two general categories of leadership style—task-oriented leaders and people-oriented leaders. A **task-oriented leader** is highly concerned about the team's purpose and task at hand. He likes to plan, carefully define the work, assign specific task responsibilities, set clear work standards, urge task completion, and carefully monitor results. A **people-oriented leader** is warm and supportive toward team members, develops rapport with the team, respects the feelings of followers, is sensitive to followers' needs, and shows trust in followers.

For a team, is one style of leadership better than another? A successful team needs both styles of leadership in some proportion. Earlier it was stated that a team has a purpose and a personality. The team must be task-oriented, but it also must cultivate a team personality and a positive environment in which team performance can flourish. This is further evidence for the need for individual team members to provide leadership that complements the leadership of the appointed leader. If the appointed leader tends to be more task-oriented, the team can balance this with a people-orientation. If the appointed leader is people-oriented, he will likely need individuals on the team to assist in details of task management. The best leaders welcome complementary leadership from the teams.

7.6   **Team Organizational Structures**

Some engineering design teams are asked to create an organizational chart. Preparing this chart gives the team the opportunity to define roles and to give thought to the relationship between the team and the team leader. Most student teams, in part because they don't know any better, default to a traditional team structure such as is given in the first block in the figure below. This structure is easy to draw, but it may or may not accurately represent the best way the team should operate. It implies a strong leader who largely directs the actions of the group, possibly with little participation or discussion from the other team members. It suggests separation between the leader and the other team members.

The second structure portrays a participative leadership/team model in which the leader is positioned in close proximity, with short, direct communication paths to all of the members. The figure implies direct accountability of the leader to all of the members and dependence of the leader on the participation of all the members.

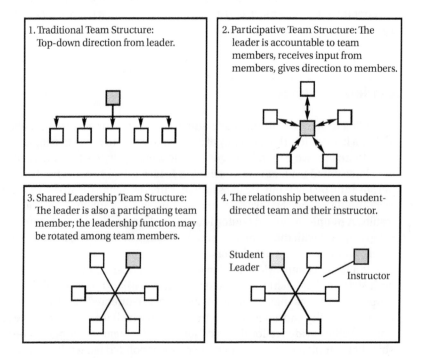

Figure 7.2 Possible team structures.

The third structure is similar to the second except it emphasizes the leader's role as a working member of the team. It suggests a flat structure where the leader is an equal rather than a hierarchical structure where the leader is above the team. For this structure it is easy to visualize the leadership function shifting around the ring from member to member as situations arise needing the special expertise of individual members.

The fourth structure shows the relationship between a student team and its instructor. The student team may be a design group, a research team, or a collaborative study unit. The instructor, though not part of the team, will be nearby and will serve as an important resource to the team. He may be asked to advise the team on administrative issues, to act as a technical consultant, or to assist in intervention or disciplinary action for a non-performing team member. All teams—in industry, in the university, and in private organizations—ultimately operate under the authority of a leader or administrator.

There is no one "right" structure. None of these structures is inherently good or evil. Examining these example structures gives the team the opportunity to define roles and to give thought to the relationship between the team and the team leader. A team needs to choose a structure that effectively models how they want to work as a team. The ultimate success of the team then hinges upon the team and its leader(s) working together to accomplish the team's purpose. As a team works, they may find that their operational structure may shift periodically in response to different situations and challenges that they face.

7.7 **Effective Decision Making**

Every team makes decisions in order to accomplish its work. The ability to make high quality decisions in a timely manner is a mark of a great team. Not surprisingly, the *process* of how decisions are made greatly affects the *quality* of these decisions. Unfortunately, most teams arrive at decisions without even knowing how they did it. In this section, we present a variety of ways in which decisions are typically made and discuss their advantages and disadvantages. Though it is recommended that a team decide in advance what decision-making process it will use, the best teams use a variety of different means of making decisions depending on the decision circumstances. The following classification of ways teams make decisions is summarized from a section in the book, *Why Teams Don't Work*, by Harvey Robbins and Michael Finley.

Consensus. A decision by consensus is a decision in which all the team members find common ground. It does not necessarily mean a unanimous vote, but it does mean that everyone had an opportunity to express their views and to hear the views of others. The process of open sharing of ideas often leads to better, more creative solutions. Unfortunately, achieving consensus can take much time and become unwieldy for larger teams.

Majority. Another way to make a decision is by majority vote. Take a vote and whatever receives the most votes wins. The advantage of this approach is that it takes less time than reaching consensus. Its disadvantage is that it provides for less creative dialogue than consensus, and there will always be a faction—the minority—who lose the vote and may become alienated.

Minority. Sometimes a small subset of the team—for example, a subcommittee—makes the decision. The advantage of this is that it may expedite the decision. Its disadvantage is that there is less overall team communication, and some team members may be prevented from making contributions to the decision.

Averaging. Averaging is compromise in its worst form; it's the way Congress and some committees arrive at decisions. Averaging is often accomplished with haggling, bargaining, cajoling, and manipulation. Usually no one is happy except the moderates. The advantage of averaging is that the extreme opinions tend to cancel one another out. The disadvantages of this are that there is often little productive discussion (as in consensus), and that the least informed can cancel the votes of the knowledgeable.

Expert. When facing a difficult decision, there is no substitute for expertise. If there is an expert on the team, he/she may be asked to make a decision. If a team lacks an expert on a particular issue, which is often the case, the best teams recognize this and seek the advice of an expert. The advantage of this is that theoretically, the decision is made with accurate, expert knowledge. The disadvantage of this is that it is possible to locate two experts who, when given the same information, disagree on the best course of action. How is a team assured that the expert gives them the best advice? Team members may be divided on which expert to consult, as well as on their assessment of the expert's credentials.

Authority Rule without Discussion. This occurs when there is a strong leader who makes decisions without discussing the details with or seeking advice from the team. This works well with small decisions, particularly of an administrative nature, with decisions that must be made quickly, and with decisions that the team is not well qualified to contribute to, anyway. There are many disadvantages to this approach, however. The greatest is that the team's trust in its leader will be undermined. They will perceive that the leader doesn't trust them and is trying to circumvent them. If one person is continually taking it upon himself to make all the decisions, or if the team is abrogating its responsibility to act together and by default, forcing the decision onto an individual, then this is a group, not a team. It is a loose aggregation of individuals, a dysfunctional team.

Authority Rule with Discussion. This can be a very effective way of making decisions. It is understood from the outset that the final decision will be made by one person—the leader or a delegated decision-maker—but the leader first seeks team input. The team meets and discusses the issue. Many, but perhaps not all, viewpoints are heard. Now, more fully informed, the appointed individual makes the decision. The advantage of this method is that the team members, being made part of the process, feel valued. They are more likely to be committed to the result. This type of decision-making process requires a leader and a team with excellent communication skills, and a leader who is willing to make decisions.

7.8   Attitudes Toward Team Experiences

1. Determine to give your best to help the team grow as a team and to accomplish its purpose. Don't just tolerate the team experience.

2. Do not expect to have perfect teammates. You're not perfect; neither are they. Don't fret that your friends are not on your team. Make friends with those on your team.

3. Be careful about first impressions of your team. Some team members may seem to know all the right things to say, but you may never see them again. Others might at first look like losers, but in the long run they are the ones who always attend meetings, make steady contributions, and ensure the success of the team. Be careful about selecting a leader at the very first team meeting. Sometimes the person who seems at first to be the most qualified to lead will never make the commitment necessary to be the leader. Look for commitment in a leader.

4. Be a leader. If you are not the appointed leader, give your support to the leader and lead from your position on the team. Don't be scared off by team experiences in the past where the leader did all the work. Be a leader who encourages and grows other leaders.

5. Help the team achieve its own unique identity and personality. A team is a kind of corporation, a living entity with special chemistry and personality. No team is just like any other team. Every team is formed at a unique time and setting, with different people and a different purpose. Being part of a productive team that sparkles with energy, personality and enthusiasm is one of the rewards of teamwork. Being part of a great team is a great memory.

6. Be patient. Foster team growth, and give it time to grow. Teammates are at first aliens, unsure of one another and of the team's purpose. Watch for and help the team grow through the stages of Forming, Storming, Norming, and Performing.

7. Evaluate and grade yourself and your team's performance. Document its successes and failures. Dedicate yourself to be one who understands teams and who acts as a catalyst, one who helps the team perform at a high level. Note team experiences in your undergraduate portfolio, especially times when you took a leadership role. Prepare to communicate these experiences to prospective employers.

7.9 Documenting Team Performance

Gain the Most from Team Experiences by Critical Evaluation

The primary aim of this chapter has been to give you tools and insights into teams that will significantly increase the probability of you being part of (and being an active leader in) great teams. You will likely be involved in many teams during your college years, in your professional life, and in volunteer capacities. Teams are everywhere, but few know how to make teams work. For teams, Voltaire's words are applicable: "Common sense is not so common." If you work hard to be uncommonly sensible about teams, you will be of great value to an employer. Approach every team experience as an opportunity to learn and to grow as a team member and leader. An excellent way to gain the most benefit is to critically evaluate each team experience. Some suggested dimensions for evaluating teams are as follows:

1. Did the team accomplish its purpose? Did it get the job done?
2. Did it do the job well? Were the results of high quality? If not, why?
3. Did the team grow through all of the developmental stages (Forming, Storming, Norming, Performing)? Were there any detours?
4. Reflect on the team's personality. Did the team enjoy working together? Did team members inspire one another to greater creativity and energy?
5. Evaluate team members on a report card like the one shown in Table 7.2.
6. Evaluate the team leader(s). Was he/she an effective leader? Why or why not? How could he/she be effective?
7. Honestly evaluate your contribution to the team.

Report Team Evaluations to Supervisor

In addition to learning more about teams, another reason to document team performance is to inform your instructor (when you are in college) or your supervisor (when at work) about the team's work. Virtually all teams have a supervisor to whom the team is accountable. It is important that you report data to them. Some argue that members of student teams should all get the same grade; in general, this is the best practice.

Table 7.2 Team Member Report Card

Criteria	Team Member				
	Ahmad	George	Jaime	Joe	Sally
1. Attendance: Attends all team meetings, arriving on-time or early. Dependable, faithful, reliable.					
2. Responsible: Gladly accepts work and gets it done. Spirit of excellence.					
3. Has abilities the team needs. Makes the most of abilities. Gives fully, doesn't hold back. Communicates.					
4. Creative. Energetic. Brings energy, excitement to team. "Can do" attitude. Sparks creativity in others.					
5. Personality: Positive attitudes, encourages others. Seeks consensus. Fun. Brings out best in others. Peacemaker. Water, not gasoline, on fires.					

Average Grade:
Grading Scale: 5 - Always; 4 - Mostly; 3 - Sometimes; 2 - Rarely; 1 – Never

But students should know from the start that if they do not participate in a significant way in the team's work, then their instructor may adjust downward the individual's team grade. In order for the instructor to make such determinations, it greatly helps if he is given (both at midterm and at the end of the semester) quantitative information on team member contributions. Sometimes instructors ask team members to grade one another, but the reasons for the grades remain unclear. The act of team members evaluating one another on specific team success criteria gives concrete evidence for team member grades. These criteria are even more valuable to the team if they are held up as a target or goal at the beginning of the team activity and if it is made clear that they will be the basis for evaluating the team.

In addition to the team member report card, the following should be valuable to your instructor:

1. Attendance records for all team meetings (usually weekly). Who attended? Was anyone habitually late or absent?
2. Actual contributions of each member. It works well to include in the team's final report a table like Table 7.3, noting the percentage contribution of each individual on each of the team's major projects/assignments. Sum these percentages in order to quantify a team member's total contribution. Variation is expected in the contributions of the different members because it is impossible to exactly balance effort. In the example below, four members' contributions varied from 95 to 155 percent. Also included in the table, and a sign of a healthy team overall, is the fact that each of the four active team members took a leadership role on at least one aspect of the project effort: Sally took primary lead on the final report, George led in the creation of the Instruction Manual, Ahmad led in the AutoCAD Drawings and the PowerPoint presentation, and Jaime led in the project construction. Though their contribution percentages vary, there is little evidence that their grades should be adjusted downward; they each deserve, it appears, the team grade. Joe, on the other hand, contributed nothing, and deserves a grade penalty. Chances are, the 10 percent is a gift from his teammates.

Table 7.3 Team Member Contributions on Primary Semester Projects

Team Members	AutoCAD Drawings	Project Construction	Instruction Manual	Final Report	Power Point Presentation	Total Indiv. Contribution
Sally	0%	15%	45%	60%	15%	**135%**
Joe	0%	5%	0%	5%	0%	**10%**
George	10%	15%	55%	15%	10%	**105%**
Ahmad	75%	10%	0%	10%	60%	**155%**
Jaime	15%	55%	0%	10%	15%	**95%**
	100%	**100%**	**100%**	**100%**	**100%**	**500%**

REFERENCES

Bennis, Warren, and Biederman, Patricia Ward, *Organizing Genius: The Secrets of Creative Collaboration*, Addison-Wesley, 1997.

Katzenbach, Jon R., and Smith, Douglas K., *The Wisdom of Teams: Creating the High-Performance Organization*, HarperBusiness, 1993.

Nuese, Charles J., *Building the Right Things Right: A New Model for Product and Technology Development*, Quality Resources, 1995.

Robbins, Harvey A., and Finley, Michael, *Why Teams Don't Work: What Went Wrong and How to Make it Right*, Peterson's/Pacesetter Books, Princeton, New Jersey, 1995.

Schermerhorn Jr., John R., *Management*, Fifth Edition, John Wiley and Sons, 1996.

Scholtes, Peter R., Joiner, Brian L., and Streibel, Barbara J., *The TEAM Handbook*, Second Edition, Oriel Incorporated, 1996.

Schrage, Michael, *Shared Minds: The New Technologies of Collaboration*, Random House, New York, 1990.

Tuckman, Bruce W., "Developmental Sequence in Small Groups," *Psychological Bulletin*, vol. 63 (1965), pp. 384–389.

Ullman, David G., *The Mechanical Design Process*, Second Edition, McGraw-Hill, 1997.

Whetten, David A., and Cameron, Kim S., *Developing Management Skills*, Third Edition, Harper Collins, 1995.

EXERCISES AND ACTIVITIES

7.1   Some individuals consider teams as the answer to all human and organizational problems. Conversely, some individuals asked to participate on teams think that being part of a team is a pain. Comment on these propositions.

7.2   From your perspective, is it easy or hard for humans to work together as teams? Does it come naturally, like falling out of bed, or is a lot of work involved in making a great team?

7.3   Describe in your own words the growth stages of a team. Give at least two examples from your own team development experiences. Have you been a part of a team that developed to a certain point and stalled? A team that progressed quickly to a productive, performing team? A team that experienced prolonged rocky periods but ultimately made it to the performing stage? If possible, note factors that led to the success and failure of these teams.

7.4   With a team of three to five people, role play a sample of conversation(s) and team behaviors for each of the five team growth stages: Forming, Storming, Norming, Performing, Adjourning.

7.5   With a team of three to five people, role play a sample of conversation(s) and team behaviors for teams making decisions by consensus and three other decision-making methods given in the chapter.

7.6 There are two quotes (source unknown) based on the T.E.A.M acronym: "Together Excellence, Apart Mediocrity," and "Together Everyone Achieves Much." These communicate the message that by working together humans can achieve more than what they could achieve working alone. Other quotes related to this are "The whole is greater than the sum of the parts," and "None of us is as smart as all of us." Do you believe that this is true? Why, or why not? Does it depend on the situation? Are some problems more suited to teams? Give examples.

7.7 Look up the word "synergy" in a college dictionary and give its basic meaning and its root meaning from the Greek. In an introductory biology textbook look up examples of "synergism." Write and briefly describe three examples of synergism from the biological world. Discuss lessons that humans can learn from these biological examples.

7.8 There are a number of historical examples where individuals have collaborated and accomplished what likely was not possible if they acted alone. Have you ever been part of a fruitful collaboration? If so, what? Why was it fruitful? Or have you been part of a failed collaboration? If so, why?

CHAPTER 8

Engineering Design

8.1 What Is Engineering Design?

Engineers create things. Engineers build things. In order to successfully perform these tasks, engineers must be involved in design or in a design process. So what is engineering *design* and an engineering *design process*? Webster's dictionary defines design as "to create, fashion, execute, or construct according to plan." It defines a natural process as "a natural phenomenon marked by gradual changes that lead toward a particular result: a series of actions or operations conducing to an end; a continuous operation or treatment, especially in manufacture."

This chapter may seem closely related to Chapter 5, "Problem Solving." Indeed, there are many similar topics and themes in Chapter 5 that are important in Design. One of the critical areas which impacts Design is the issue of "external constraints" that can influence the outcome of the process. Appropriate references back to Chapter 5 will be noted throughout this chapter.

In the work of engineers, engineering design is an important and ongoing activity. Students who graduate with an engineering degree from an accredited program have had a significant amount of design experience as part of their education. The Accreditation Board for Engineering and Technology (ABET) has traditionally defined engineering design as follows:

> *Engineering design is the process of devising a system, component, or process to meet desired needs. It is a decision-making process . . . in which the basic sciences and mathematics and engineering sciences are applied to convert resources optimally to meet a stated objective. Among the fundamental elements of the design process are the establishment of objectives and criteria, synthesis, analysis, construction, testing, and evaluation. . . . it is essential to include a variety of realistic constraints, such as economic factors, safety, reliability, aesthetics, ethics, and social impact.*

ABET then describes the basic educational components for a major engineering design experience that must be included in an engineering program:

Each educational program must include a meaningful, major engineering design experience that builds upon the fundamental concepts of mathematics, basic sciences, the humanities and social sciences, engineering topics, and communication skills. The scope of the design experience within a program should match the requirements of practice within that discipline. . . . all design work should not be done in isolation by individual students; team efforts are encouraged where appropriate. Design cannot be taught in one course; it is an experience that must grow with the student's development. A meaningful, major design experience means that, at some point when the student's academic development is nearly complete, there should be a design experience that both focuses the student's attention on professional practice and is drawn from past course work. Inevitably, this means a course, or a project, or a thesis that focuses upon design. "Meaningful" implies that the design experience is significant within the student's major and that it draws upon previous course work, but not necessarily upon every course taken by the student.

Many entering engineering students often confuse the "design process" with drafting or art-related work. To clarify this, ABET adds the following:

Course work devoted to developing computer drafting skills may not be used to satisfy the engineering design requirement.

Most engineering programs concentrate their engineering design coursework in the later part of the student's program, thereby allowing the student to apply much of the prerequisite background in math, science and related fields to the various engineering problems. In most engineering programs, design courses comprise between 20% and 25% of the total curriculum.

8.2 **The Design Process**

It is important to realize that there is not one uniform approach to engineering design that is followed by practicing engineers. Some firms approach engineering design as a short, simple process with only a few steps, while others use a more complex, multi-step method with several stages. No matter what process is used, it is important to realize that engineering design is always continuous. The completion of one design, or the solution of one problem, may serve to open up opportunities for subsequent designs or modifications.

A design process is used whether a product is being developed for an ongoing manufacturing process, where thousands or even millions of a certain item will be produced, or for a one-time design, such as with the construction of a bridge, dam, or highway exit ramp.

10-Stage Design Process

This chapter will present one design process (the 10-stage process) that is presently in use. This process will then be applied to an existing product to analyze how each of the 10 stages may have been applied in the product's realization.

The 10 stages that make up the process are as follows:

- Stage 1: Identify the problem/product innovation
- Stage 2: Define the working criteria/goals
- Stage 3: Research and gather data
- Stage 4: Brainstorm/generate creative ideas
- Stage 5: Analyze potential solutions
- Stage 6: Develop and test models
- Stage 7: Make the decision
- Stage 8: Communicate and specify
- Stage 9: Implement and commercialize
- Stage 10: Perform post-implementation review and assessment

The actual process begins with the appointment of a project manager or team leader. This individual will be responsible for oversight of the entire process to ensure that certain key elements of each stage have been satisfied before the project moves on to the next stage. This person will also be responsible for recruiting team members of varying backgrounds and expertise for each of the stages. The team will include non-engineers as well as engineers. Some of the team members will be used throughout the process, while others will be needed only for certain parts of the process.

Stage 1: Identify the Problem

Engineers are problem-solvers; and the problems they solve are often identified as the specific needs and problems of customers. For example, a new prosthesis may be required to overcome a particular handicap, or increased gas mileage standards demand higher-efficiency engines, or a new computer program is needed to monitor a modified manufacturing process, or new safety devices are required to better protect infants in automobiles. Therefore, the first stage to problem solving in engineering design is to establish the actual problem clearly, and to identify sources of information to help understand the scope and nature of the problem.

The project manager will call upon the resources of various individuals to assist with these initial stages in the process. There may be a multitude of sources outside as well as within the organization that can assist with solving the problem. Many firms have a research and development unit made up of scientists and engineers who possess the training and expertise to assist with problem evaluation. In addition, sales engineers, who maintain consistent contact with outside individuals, can provide valuable input on problem identification. If the problem is one of expanding a current product line or modifying an existing system to improve it or make it better fit customer needs, management will likely play a role in the definition of the problem. Each group would be represented on the team at this stage.

External resources may come from trade shows, conferences, technical presentations, patent listings, and publications. Ideas generated from existing or prospective clients may also be valuable. Information gathered from external research agencies, private laboratories, and government-funded foundations can be useful as well. In addition, an awareness of the competition and what products and services they are involved with can be beneficial in problem identification. At this early stage in the process, it is helpful to establish a preliminary, formal statement of the problem. For example, the need for a new automobile safety device for infants might evolve into a preliminary statement of the problem as follows: "Develop a better child restraint system that will protect children involved in automobile collisions."

It is important at this point to review the section of Chapter 5 which discusses the "real Problem." Before going ahead with the design process, the team must be certain that the correct issues and background have been thoroughly explored so the most appropriate solutions can be developed.

At this point, if the project manager is satisfied that all necessary issues have been resolved, the project will likely be passed on to the next stage where project criteria and goals will be developed.

Stage 2: Define the Working Criteria and Goals

Once the problem has been identified, it is important for the team to be able to validate it throughout the design process. This requires the establishment of certain working criteria, or standards, which can be used in each of the 10 stages to measure possible solutions. The ultimate objective in this stage is to be able to establish preliminary goals which will act as the focal point of the team as it works through the process. The development of some working criteria provides a means to compare possible solutions. At this stage of the design process, everything is preliminary, so it is still possible for the team to modify the criteria if necessary.

Examples of working criteria could include answers to the following questions:

1. How much will it cost?
2. Will it be difficult to produce?
3. What will be the size, weight, strength?
4. What will it look like?
5. Will it be easy to use?
6. Will it be safe to use?
7. Are there any legal concerns?
8. Will it be reliable and durable?
9. Can it be recycled?
10. Is this what the customer truly wanted?

11. Will our customers want to purchase it?
12. Will customers want to purchase this version instead of a competitor's product?

Once some preliminary working criteria have been established, it is important to try to develop overall goals for the process. This is basically a statement of objectives which can be evaluated as the design process evolves. Using the example of new standards implemented to increase gas mileage and reduce emissions in automobiles, the goals for the design might be: "To develop an automobile engine which produces 25% less emissions while increasing gas mileage by 10%."

Having overall goals established for the project provides a means of evaluating, monitoring, and changing, if necessary, the focus of the process as it evolves through the 10 stages. For the project manager, the criteria and goals become a "checkpoint" for assessing the progress to date, and will help them determine if the project is ready to move to the next stage, or if the process needs to return to Stage 1 for re-evaluation.

Stage 3: Research and Gather Data

This stage is very important to all remaining stages of the design process. Having good, reliable background information is necessary for the team to begin exploring all relevant aspects of the problem. Consistent with the preliminary working criteria and the goals that have been established, the team members selected for this phase of the process must determine what types of information will be needed and the best sources of that information. For example, they may want to know such things as:

1. What information has been published about the problem?
2. Is there a solution to the problem that may already be available?
3. If the answer to the above is yes, who is providing it?
4. What are the advantages of their solution?
5. What are the disadvantages of their solution?
6. What is the cost?
7. Is cost a significant issue?
8. What is the ratio of time spent compared to overall costs?
9. Are there legal issues to consider?
10. Are there environmental concerns which must be considered?

There are many possible resources the team members can utilize to assist them in their research. A good starting point may be a simple search using the Internet. This may provide useful sources of material that can serve as the focus for additional research. Other sources of reference information may come from:

1. libraries
2. professional associations (technical and non-technical)
3. trade journals and publications
4. newspapers and magazines

5. market assessment surveys
6. government publications
7. patent searches and listings (U.S. Patent Office: www.uspto.gov.)
8. technical salespersons and their reference catalogs
9. professional experts including engineers, professors, and other scientists
10. the competition's product (How do they construct it? Disassemble their product and study it.)

Detailed files, notes, pictures, sketches, and other supporting materials will be maintained by the team to assist them as they proceed through the remaining stages of the design process. As new supporting information is discovered by the team, the material will be added as additional reference resources.

Depending on the type of information that has been collected, at this point it may be appropriate to review the established preliminary working criteria and the overall goals. To assist the project manager throughout the process, some modifications may need to be made. For example, it may be determined that one or more of the criteria may not actually apply to the problem. Likewise, new issues may surface which may necessitate the addition of new criteria or a modification of the goals. It is important to have these issues resolved before moving on to the next stage.

Stage 4: Brainstorm/Generate Creative Ideas

The basic concept involved at this stage of the process is to creatively develop as many potential solutions to the problem as possible. The more ideas that can be generated, the better the likelihood of identifying a feasible solution to the problem. The project manager will want to gather a group of individuals from both technical and non-technical backgrounds to provide their unique perspective to the problem. This group may include engineers, scientists, technicians, shop workers, production staff, finance personnel, managers, computer specialists, and perhaps even a few clients.

A major method of generating multiple ideas to a problem is called *Creative Problem Solving*, using a technique called *brainstorming*. (Note: For a thorough discussion of creative problem solving and brainstorming strategies, please refer to Section 5.6 in Chapter 5.) With this method, a large group of individuals with varying backgrounds and training are brought together to attempt to solve a particular problem. Every idea that is spontaneously contributed from the group is recorded. The basic premise is that no idea is deemed too wild or illogical at this stage. No preliminary judgments are made about any member's idea, and no negative comments are allowed. The goal is to develop a long list of possible alternative solutions to the problem at hand. The group leader should be able to encourage participants to suggest random thoughts and ideas.

Some students may have had the opportunity to engage in brainstorming exercises. Brainstorming can be fun and highly stimulating to the creative process. For example: How many ways can you suggest to use a piece of string and a Styrofoam drinking cup? What could be created from a trash bag that contains some old magazines, tape and a ruler? Think of ways your student organization could earn extra funds for a field trip, etc.

It is conceivable that a brainstorming session would be continued on a second occasion to allow members time to consider other possible options. When the group reconvenes, members may have several new ideas or new perspectives for examining the problem.

At the conclusion of this stage, the group should have a long list of potential solutions to the problem. It is important to remember that at this stage of the process no idea has been eliminated from consideration. Each idea will be evaluated eventually, but it is important to keep all options open at this stage.

Once the project manager is satisfied that all possible solutions have been suggested, the project will likely be cleared for the next stage.

Stage 5: Analyze Potential Solutions

Note: Chapter 5.3 discussed the Analytic Method in detail. That information is directly related to this material in this stage. In the early part of the analysis stage (Phase I), it is important to try to narrow the ideas generated in the brainstorming stage to a few ideas which can be subjected to more sophisticated analysis techniques. This early narrowing could include:

- Examine the list and eliminate duplicates. As discussed earlier, it is important not to create limited categories, but only to eliminate repeated ideas. If two are similar, both should remain at this point.
- Allow the group to ask clarifying questions. This could help identify duplicate ideas.
- Ask the group to evaluate the ideas. The group members can vote for their top three ideas, and those that gain the most votes will be retained for more detailed analysis.

At this point, there will probably be a small number of ideas remaining. These can now be analyzed using more technical and perhaps time-consuming analysis techniques (Phase II).

A variety of individuals should be involved at this stage, but the engineer will be of primary importance. The analysis stage requires the engineer's time and background. It is here that one's training in mathematics, science, and general engineering principles are extensively applied to evaluate the various potential solutions. Some of the techniques in this phase can be time consuming, but a thorough and accurate analysis is important before the project moves to the next stage in the process. For example, if the problem under consideration was the development of an automobile bumper that could withstand a 20-mile-per-hour crash into a fixed object barrier, several forms of analysis could be applied, including:

Common sense: Do the results seem reasonable when evaluated in a simple form? Does the solution seem to make sense compared to the goal?

Economic analysis: Are cost factors consistent with predicted outcomes?

Analysis using basic engineering principles and laws: Do each of the proposed solutions satisfy the laws of thermodynamics? Newton's laws of motion? The basic principles of the resistance of a conductor, as in Ohm's law, etc.?

Estimation: How does the performance measure up to the predicted outcomes? If the early prediction was that some of the possible bumper solutions would perform better than others, how did they perform against the estimate? (Review Chapter 5.3 for a more thorough discussion of Estimation.)

Analysis of compatibility: Each of the possible solutions and their related mathematical and scientific principles are compared to the working criteria to determine their degree of compatibility. For example: How would each bumper solution meet the criteria of being cost-effective? What would be the size, weight, and strength of each of the proposed solutions? How easy would each one be to produce?

Computer analysis techniques: One frequently used method is *finite element analysis*. With this method, a device is programmed on a computer and then numerically analyzed in segments. These segments are then compared mathematically to other segments of the concept. In the bumper crash example, the effects of the impact could be analyzed as a head-on crash and then compared to a 45-degree angle collision or a side-impact crash. As each section is analyzed, the "worst-case scenario" can be evaluated.

Conservative Assumptions: As discussed in Chapter 5, this technique can be most useful in analysis. It can build safeguards into the analysis until more data is generated.

After each of the working criteria have been examined and compared to the list of possible solutions, a process eliminates those that have not performed well in the various forms of analysis. It is expected at this point that only three to five options from the original list of prospective solutions will remain. These remaining options will then be reviewed by the project manager, who will likely authorize that the project be cleared for the next stage, assuming that these remaining options meet the working criteria and overall goals. If not, the process will need to be terminated, or return to an earlier stage to correct the problem.

Stage 6: Develop and Test Models

Once each of the prospective solutions has been analyzed and the list of feasible options has been narrowed to a few possibilities, it is time to enter the phase where specific models will be developed and tested. Again, in this stage it is important to have a strong background in engineering coupled with experience and sound judgment. However, this stage will also involve team members who are computer specialists, shop workers, testing technicians, and data analysts.

There are several types of models which are commonly used by engineers and others in this stage. These include:

Mathematical models: Various conditions and properties can be mathematically related as functions and compared to one another. Often these models will be computerized to assist in visualizing the changing parameters in each of the models.

Computer models: There are various types of computer models which can be used. Typically these models allow the user to create on-screen images which can be analyzed prior to the construction of physical models. The most common computer modeling is referred to as CAD (Computer Aided Design) where models are designed and displayed as three-dimensional wire-frame drawings or as shaded and colored pictures. The computer can also be used to control equipment that can generate solid models using techniques such as "stereo-lithography," where quick-hardening liquids are shaped into models or other forms of "rapid prototyping." These on-screen models, or the prototype models they produce, can then be used in the testing process.

Scale models: Typically, these smaller models have been built to simulate the proposed design but may not include all of the particular features or functions. These models are often called prototypes or mock-ups and are useful in helping engineers visualize the actual product. Such models may be used to depict dams, highways, bridges, new parts and components, or perhaps the entire body of a prototype automobile.

Diagrams or graphs: These models provide a tool for visualizing, on a computer or on paper, the basic functions or features of a particular part or product. These diagrams or graphs could be the electrical circuit components of an operating unit of the product, or a visualization of how the components eventually may be assembled.

Once the models have been developed and created, it is time to test each of them. Performing a variety of tests on each of the models allows for comparison and evaluation against the working criteria and the overall goals that have been established. In actuality, tests are done continually throughout a project including early models, prototypes, and the testing of product quality as the product is manufactured or built. However, the results of the testing done in this stage establish the foundation for the decisions that will be made about the future of this project.

Examples of these tests include:

Durability: How long will the product run in testing before failure? If the product is a structure, what is its predicted life span?

Ease of assembly: How easily can it be constructed? How much labor will be required? What possible ergonomic concerns are there for the person operating the equipment or assembling the product?

Reliability: These tests are developed to characterize the reliability of the product over its life cycle, to simulate long-term use by the customer.

Strength: Under what forces or loads is a failure likely, and with what frequency is it likely to occur?

Environmental: Can the parts be recycled?

Quality consistency: Do tests show that product quality is consistent in the various stages? Is the design such that it can be consistently manufactured and assembled? What conditions need to be controlled during manufacture or construction to ensure quality?

Safety: Is it safe for consumer use?

Consistency of testing: This technique examines each of the testing methodologies against the various results obtained to determine the consistency among the various tests. In the bumper design example, the testing methodology might be different for a head-on impact test than for a 45-degree crash test in order to minimize testing inconsistencies.

All of this information will be evaluated by the project manager. If the manager is satisfied that these results consistently meet the working criteria and overall goal, the project will likely be cleared for the next stage.

Stage 7: Make the Decision

At this stage, it is important for the team members to establish a means to compare and evaluate the results from the testing stage to determine which, if any, of the possible solutions will be implemented. The working criteria that have been used throughout the process are the critical factors which will be used to determine the advantages and disadvantages of each of the remaining potential solutions. One of the ways to evaluate the advantages and disadvantages of each of the proposed solutions is to develop a *decision table* to help the team visualize the merits of each. Typically, a decision table lists the working criteria in one column. A second column assigns a *weighted available point total* for each of the criteria. The team will need to determine the order of priority of each of the criteria. The third column then provides *performance scores* for each of the possible solutions. A sample Decision Table might look like the one that follows.

TABLE 8.1 A Decision Table

Working Criteria	Points Available	#1	#2	#3
Cost	20	10	15	18
Production Difficulty	15	8	12	14
Size, Weight, Strength	5	5	4	4
Appearance	10	7	6	8
Convenient to use	5	3	4	4
Safety	10	8	7	8
Legal issues	5	4	4	4
Reliability/Durability	15	7	9	11
Recyclability	5	4	3	4
Customer Appeal	10	7	8	9
Total	**100**	**63**	**72**	**84**

Based on this sample decision table, it appears that while none of the proposed solutions have scored near the "ideal" model, Solution #3 did perform better than the others. Using this information, the project manager and team leaders would make the final decision to "go" or "no-go" with this project. They may decide to pursue Solution #3, to begin a new process, or even to scrap the entire project. Assuming they decide to pursue Solution #3, the team would prepare the appropriate information for the next stage in the process.

Stage 8: Communicate and Specify

Before a part, product or structure can be manufactured, there must be complete and thorough communication, reporting, and specification for all aspects of the item. Team member engineers, skilled craft workers, computer designers, production personnel, and other key individuals associated with the proposed project must work together at this stage to develop the appropriate materials. Such materials include detailed written reports, summaries of technical presentations and memos, relevant e-mails, diagrams, drawings and sketches, computer printouts, charts, graphs, and any other relevant and documented material. This information will be critical for those who will be involved in determining final approval for the project, as well as the group involved in the final implementation of the product. They must have total knowledge of all parts, processes, materials, facilities, components, equipment, machinery, and systems that will be involved in the manufacturing or production of the product.

Communication is an important tool throughout the design process, but especially in this stage. If team members cannot adequately sell their ideas to the rest of the organization, and be able to appropriately describe the exact details and qualities of the product or process, then many good possible solutions might be ignored. At this stage of the process, it also may be important to create training materials, operating manuals, computer programs, or other relevant resources which can be used by the sales team, the legal staff, and prospective clients and customers.

At this point, if the project manager is satisfied that all necessary materials have been adequately prepared and presented, the project will likely be passed on to the next stage, the implementation stage.

Stage 9: Implement and Commercialize

The next-to-last stage of the design process is critical, as it represents the final opportunity for revision or termination of a project. At this point in the process, costs begin to escalate dramatically, so all serious issues should be resolved by this time.

In addition to the project manager and team leaders, there are a number of other individuals involved at this stage, representing a variety of backgrounds and areas of expertise. While engineers are a part of this stage, many of the activities here are performed by others. Some of those involved in this stage may include:

Management and key supervisory personnel. These individuals will make the ultimate decisions concerning the proposed project. They are concerned with the long-term goals and objectives of the organization, determining future policies and programs that support these goals, and making the economic and personnel decisions that affect the overall health of the organization.

Technical representatives. These may include skilled craft workers, technicians, drafters, computer designers, machine operators, and others involved in manufacturing and production. This group will have primary responsibility for getting the product "out the door."

Business representatives. This group may consist of:

- **human resource personnel** if new individuals must be hired
- **financial people** to handle final budget details and financial analysis questions
- **purchasing personnel** who will procure the needed materials and supplies
- **marketing and advertising staff members** who will help promote the product
- **sales people** who will be involved in the actual selling and distribution of the product

Attorneys and legal support staff. The legal representatives who will handle a variety of legal issues including patent applications, insurance, and risk protection analysis.

If all parties are in agreement that all criteria have been satisfied and the overall goal achieved, the actual production and commercialization will begin. However, there is one remaining stage where the project activities and processes will be evaluated and reviewed. (Note: Some stages of the process may be monitored differently depending on whether the project relates to a one-time product, i.e., a bridge or dam, or an ongoing manufacturing process.)

Stage 10: Perform Post-Implementation Review and Assessment

At this point, it is assumed that the project is in full production. The project manager, key supervisory personnel, and team members who had significant input with the project are gathered together for a final project review and assessment. This stage involves the termination of the project team, since the product is now considered to be a regular product offering in the firm's overall product line. The product's performance is reviewed, including the latest data on production efficiency, quality control reports, sales, revenues, costs, expenditures, and profits. An assessment report is prepared which will detail the product's strengths and weaknesses, outline what has been learned from the overall process, and suggest ways that teams can improve the quality of the process in the future. This report will be used as reference for future project managers and teams to consult.

EXERCISES AND ACTIVITIES

8.1 Disassemble one of the devices listed below, and put it back together. Sketch a diagram of all the parts and illustrate how they fit together to make the device operate. List at least three ways you think the design could be improved. Choose one of the following devices: a) flashlight, b) lawn sprinkler, c) sink faucet, d) stapler, e) toaster, f) computer mouse.

8.2 Prepare a list of questions that should be resolved in identifying each of the following problems:

a) Develop an improved manual gearshift for a mountain bike.
b) Develop a better braking system for in-line skates.
c) Develop a recliner chair that incorporates six built-in surround sound speakers in the unit.
d) Develop a hands-free flashlight.
e) Develop a theft-proof bicycle lock.
f) Develop a baseball cap with wireless stereo headphones.
g) Develop a secure storage area on a bike for a helmet.
h) Develop a new idea to reduce overcrowded parking on campus.
i) Develop a hands-free cellphone that can be used when doing activities.
j) Develop a shockproof cellphone carrying case.

8.3 Develop a list of working criteria that could be used in deciding whether to:

a) Accept a co-op job offer from Company A or Company B.
b) Study overseas for the Fall semester or remain on campus.
c) Buy a new car or repair your old one.
d) Purchase a desktop or laptop computer.
e) Change your major or remain in engineering.
f) Purchase a new computer or upgrade your current model.
g) Live in the dormitory or lease an off-campus apartment next year.
h) Pledge a fraternity/sorority.

8.4 Identify five product, structure, or system designs you think can be improved. Pick one and write a preliminary problem statement for the engineering design process.

8.5 Using an item from your list in Exercise 8.4, develop a list of reference materials that would be used in developing possible solutions to the problem. Provide specific examples.

8.6 Get together with three other classmates and brainstorm at least 30 ways to use one of the following objects:

a) Two-foot length of string
b) Ping-pong ball

     c) Three plastic pop bottles

     d) Page of notebook paper and a 2-inch piece of tape

     e) Old telephone directory

     f) Trash bag

     g) Deck of playing cards

     h) Yo-yo

     i) Frisbee

     j) Metal coat hanger

     k) Empty plastic milk container

     l) 15 paper clips

     m) 2 tubes from paper towel rolls and 2 pages of newspaper

     n) Deflated balloon

     o) Bag of marshmallows

     p) Newspaper and 12" of masking tape

8.7   Using the information you developed in Exercise 8.4, prepare a decision table for three possible alternative solutions.

8.8   Read a current newspaper or magazine article (possibly on the Web) that describes and discusses a new product, device, or system. Prepare a three- to four-page essay which analyzes and explains, in detail, each of the stages of the design process that were probably involved in the engineering of this product or device. Apply each stage of the 10-stage design process discussed in this chapter to your product or device, and use specific examples to support your statements. Make sure you list your sources in the body of the paper or in a bibliography.

8.9   Put together a team of five students from your class. Using the 10-stage design process, have your team develop a portable, garage-like covering for a bicycle that can be stored somewhere on the bike when not in use.

8.10   Put together a team of four or five students from your class. Using the 10-stage design process, build the tallest possible tower that can support a 12-oz can of soda/pop. Materials: one roll of masking tape, one package of straws, and one can of soda/pop. Constraints: Your tower must be free standing. The base must fit on a plate. Your tower must stand for 5 seconds under load for measurement. After the design is complete, 30 minutes will be allowed for tower construction.

8.11   Put together a team of four or five students from your class. Using the 10-stage design process, build a bridge that spans three feet between supports that contact the ground. Materials: one roll of duct tape, ten feet of rope, a 40 inch by 75 inch piece of cardboard, and 14 cardboard slats. After the design is complete, 40 minutes will be allowed for tower construction and testing. The test: one of your team members must walk heel-to-toe across the bridge.

8.12   Put together a team of five students from your class. Using the 10-stage design process, have your team develop a carrying mechanism for Rollerblades and/or your shoes when one is not in use.

CHAPTER 9

Technical Communications

Engineers must possess the technical skills to complete engineering analysis, evaluation, and design. However, these skills are essentially useless if they cannot be communicated to the variety of audiences with whom engineers work. These audiences will have a wide range of technical skills and might come from backgrounds extremely different from yours. To communicate effectively as an engineer, you need to understand your audiences, the desired effects you wish to have on these audiences, and the conventions of communication these audiences will expect you to follow (as shown in Figure 9.1). In other words, you will need to employ a rhetorical strategy to make your attempts at communication with these disparate audiences successful.

The rhetorical triangle shown in Figure 9.1 is an illustration of the considerations to be made in any communication scenario. The triangle represents three questions you can ask yourself to help you to make good choices in any rhetorical context.

- *Who is my audience?* Do I know them personally? What are their attitudes? Needs? Values?
- *What is my purpose?* Am I trying to explain something? Share something? Sell something? How do I want the audience to respond?
- *What is expected of me as a communicator?* Are there conventional approaches to this task? How do I make the desired impression?

Figure 9.1 The rhetorical triangle.

Of course, these questions overlap; they are parts of an overall context for communicating. For example, is the context professional? Is it technical? What are the time constraints? Still, it is difficult to even begin composing effectively until the three questions implied by the rhetorical triangle have been answered.

Rhetoric is an often misunderstood word. It can have negative connotations, such as when people dismiss something as "political rhetoric" or when they think of a rhetorical question merely as something that doesn't require an answer. However, rhetoric is all around us, and it is important to all of us. Anything that explains, persuades, or motivates is a form of rhetoric. The lessons of your teachers, the advertisements in various media, and the words and images on your t-shirt are all forms of rhetoric.

You probably make many rhetorical decisions in your daily communication without recognizing them as forms of rhetoric. You would likely wear a suit to a job interview because you are concerned with the impression you will make. You know the conventions for professionalism in your field and dress appropriately. This is a decision about how you will present yourself visually to a specific audience, the interviewer, for a specific purpose, to get a job. This decision implies a strategy involving visual rhetoric, although you would likely never use those terms to describe it.

In engineering, you will be communicating with people who are making decisions concerning your work. These decision makers have minimal time, and you want to have a maximal impact. This situation means you should try to keep your communications brief and to the point. Words often have the least impact; instead, information can be communicated with pictures, charts, and graphs. Whereas other types of communication place great emphasis on complex language and nuance (which require more words to explain), technical communication usually requires you to get to the point quickly and simply.

> As we get started, here are three golden rules of communication for engineers:
>
> 1. **KEEP IT BRIEF!**
> 2. **KEEP IT SIMPLE!**
> 3. **SHOW IT!**

This chapter will help you to understand the three essential elements of effective communication (audience, effect, and conventions) and give tips on communicating in visual, spoken, and written modes.

9.1   Visual Communication

Later sections in this chapter deal with issues involving images in specific media. First, though, there are a few basic principles of visual communication that apply to all media. This section touches on a few basics that everyone should know, including graphics, image composition, color compatibility, and you.

- *Graphics*—A graph or chart that looks acceptable in your technical report might not work as well in a presentation. Posters and Microsoft PowerPoint slides have to be understood more quickly than those in reports and other written documents because the audience usually does not control the pace of a presentation. For this reason, it is a bad idea to simply accept charts and graphs in the form that programs like Microsoft Excel produce by default (see Figure 9.2). Always make sure that labels and numbers on axes of graphs are in fonts that can be easily read. You might also want to remove unnecessary lines from charts or move legends onto the chart itself to enlarge the display of data, as shown in Figure 9.3. The generic chart (Figure 9.2) might be acceptable for a report if it is properly captioned

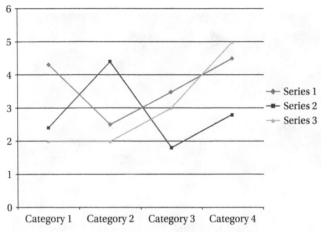

Figure 9.2 A generic chart generated by Microsoft Excel's default settings.

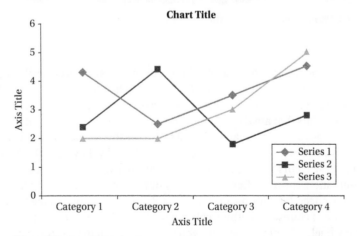

Figure 9.3 An edited chart.

and discussed in the report's text. Yet, changes to the layout will make the same information more readable in a technical presentation.

- *Image composition*—Many people assume that when you compose a picture, you place the subject of the photograph in the center. This might be true in some cases, such as when you are merely showing an item as part of a slideshow. However, artists know that photographic subjects are far more striking when they are slightly off center. The guide generally used for this approach is called the *rule of thirds*. Imagine a tic-tac-toe board laid over top of your image, as shown in Figure 9.4. Major components of that image should fall on those lines with the most powerful aspects of the picture at the intersections.

Figure 9.4 Photograph illustrating the Rule of Thirds.
(Source: Teresa Bowles)

- *Color compatibility*—The use of color in communication is critically important. A little basic understanding of color theory can help you a lot with color choices. If you look at a color wheel on this book's accompanying website (Figure 9.5), you'll see that colors that are directly opposite each other are called *complementary colors*. These are colors that work well together, and include some very well-known combinations such as red and green, or purple and yellow (gold). However, when using these colors in a light-driven medium, such as a website or a PowerPoint presentation, the font often will actually hurt the eyes of your audience due to a strobe effect that only occurs with complementary colors. In these cases you can instead choose something called a *split complement*. To do this, look at your color wheel, select the base color you know you want to use, and then draw either an equilateral or isosceles triangle with that original color as its topmost

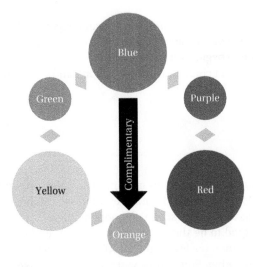

Figure 9.5 The color wheel with an illustration of complementary colors.

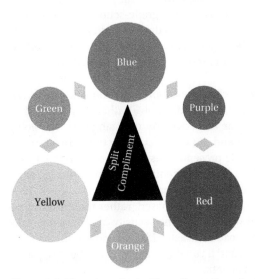

Figure 9.6 The color wheel with an illustration of split complementary colors.

vertex, as shown in Figure 9.6. The angles that form the base will point to split complementary colors.

- *You*—You communicate visually through your clothing, your posture, and your expression. Remember that an audience begins taking visual cues about you from the moment they can see you, perhaps even before you begin your presentation.

If you run around with a panicked look on your face and papers falling out of your briefcase as you try to load your presentation onto the computer in the last minute before you are scheduled to begin, you will not fill your audience with confidence in your message. Look and act confident and competent, and your audience will be more open to whatever you have to say.

9.2   Oral Presentations

The term *oral presentations* usually refers to multimodal, slide-supported presentations, often composed in Microsoft PowerPoint. Presentations can be made and supported in many ways, and as technology advances, more methods are likely to appear.

Informal presentations tend to be shorter than formal ones, and do not require as much preparation time or formal business attire. You do not usually need to prepare a PowerPoint presentation, either. Sometimes you will have to make impromptu or spontaneous presentations. You might make such presentations in a meeting with your community partner or at a meeting involving a design team.

Posters are another means of supporting oral presentations. Many students are surprised to hear that posters exist beyond grade-school book reports, yet posters can be an effective tool for sharing ideas with a diverse audience. A relatively new variation of this sort is the single-slide presentation, where a talk is supported by what is essentially a poster in digital form, but instead of being presented on an easel, it is projected on a screen.

Another example, Prezi, is cloud-based presentation software offering a nonlinear alternative to the basic PowerPoint slide show. Rather than going through a stack of slides, the presenter guides the audience around a single canvas, going into depth in specific areas by zooming in and out on different sections. This format can be a compelling method for single-slide presentations. On the other hand, this format can also invite new problems, such as drawing attention to the new format, and therefore away from the content. The nonlinear nature can also be confusing if the presentation is poorly organized. Because Prezi is cloud based, rather than residing on your computer's hard drive, some also worry about the security of proprietary information. Prezi can be useful, but it can also lead to disaster. Like any new tool, be sure you have a need for the tool before you use it, and if you do decide to use it, be sure you become familiar with its pitfalls as well as its benefits.

Poster Presentations

Traditional poster presentations are audience-driven forms of communication. Poster presentations usually feature a number of posters displayed at the same time in a presentation hall. The crowd will walk among the posters, individuals stopping when a title catches his or her eye, at which time he or she can engage in a one-on-one discussion with the presenter. For this reason, it might be effective to think of the poster as a visual abstract. It should cover the major points of a project, but should not go into the sort of detail one would expect from a written document. Often, the most difficult part of poster composition is paring down the details of a complex project to fit the limited space of a poster.

Poster Composition

Posters are not meant to display every detail of a project; rather, they should be a series of conversation starters and supporting visuals that tell the story of your project. If you try to present too much information on a poster, the audience will find the poster intimidating and impenetrable. The following are tips to help you compose a better technical poster:

- *Titles*—Keep them brief and descriptive. Many times, the titles of technical papers are very long and contain all sorts of jargon. Long titles are attention killers on a poster. Remember that a title should grab the audience members' attention as they pass. Once you have their attention, then you can begin to explain the poster's content.
- *Images*—Limit the number of images, and make sure they are relevant. Some rhetoricians recommend a limit of five to seven images per poster; however, it is difficult to give a range because pictures and posters both come in a variety of sizes. Four 5 × 5 inch pictures on a 2 × 3 foot poster might look cluttered, especially if there is a lot of accompanying text. The same number of 5 × 5 inch pictures and text on a 3 × 4 foot poster might not look cluttered at all. Use your best judgment to determine if you have too many images. It is a good idea to caption each image, as well as to number them, so that they can be referred to in the poster text.
- *Image resolution*—Raster images (e.g., photographs) taken from the web usually have a resolution of 72 pixels per inch (dpi). The standard for printed images is 300 dpi. This is why enlarged pictures sometimes become blurry. If you want to use a 5 × 5 inch photograph, you need to make sure that it has a high enough resolution to look sharp at that size: (5 × 300) × (5 × 300). Images with a resolution of 1500 × 1500 or higher will be sufficient for a 5 × 5 inch photograph.
- *Fonts*—Make sure that your fonts can be easily read from a distance of six to eight feet. The easiest way to do this is to locate the smallest font on your poster, zoom in to 100 percent view on the computer screen, and then step back six feet. If you can read your font, then your font sizes are sufficient. It is also wise to avoid novelty fonts because many of them are difficult to read and some could be perceived as unprofessional.
- *Charts, tables, equations, and graphs*—Like images, these elements should be limited in number. Also as with images, it is difficult to give a specific range of the number of images to use. Use your best judgment to ensure that the poster does not look cluttered. Charts, tables, equations, and graphs should also be thoroughly captioned and numbered.
  - Be careful using equations and tables. Although these are clearly important for understanding concepts and organizing data, they are not always the best

way to present information. In the same way that a picture is said to be worth a thousand words, a graph is worth a thousand equations, and a chart is worth a thousand tables. Audiences tend to understand things shown visually more easily than they follow the logic of an equation or table. Whenever you have the option to explain a concept using an image, a chart, or a graph, you probably should!

- *Color*—Color can help organize your poster, or it can be a distraction. Background colors can bring unity to portions of a poster, and they can imply groupings of similar information. However, colors can also have connotations that are not helpful to you. For example, using your school colors from college might reveal your pride in your school, but it might also make your audience think of you as a student when it would be to your benefit to have them think of you as a mature professional.
  - It is also important to be aware that colors have connotations, and sometimes their connotations can shift over time. For example, if you ask members of the younger generation what they think of when they see the color green, they will probably mention environmental issues. In fact, the phrase "going green" is often meant to imply that environmental concerns are being taken into account. Conversely, to older audiences, green might carry very different connotations, implying greed or perhaps envy. They might think of being "green with envy" or think of earning some "greenbacks."
  - Usually, when audience members see colors on a chart or a graph, they will try to assign a meaning to that color. Often, that meaning is intentional, and is explained through a color key somewhere on the chart itself. However, because poster contents sometimes come from multiple sources and perhaps multiple authors, it is important to be consistent with color denotations throughout your poster. Whatever red might mean on the first graph we see, its meaning should not change as we proceed to the next charts and graphs. Be consistent with color!
- *Layout*—In Western cultures, readers usually begin reading in the upper left corner, moving downward through columns the same way one would read a newspaper.
  - Posters are not newspapers, so they aren't always composed in columns. If you deviate from this pattern, you'll need to include numbers or arrows or some similar clue to help the reader know how their gaze should move over the poster.
  - Composing in columns is an established convention for poster layout. Figure 9.7 illustrates how information is usually presented in three or four vertical columns if the poster is meant to be viewed in a portrait orientation (where the height exceeds the width).
- *Rhetorical triangle*—As always, ask yourself the three questions about audience, purpose, and conventions implied by the rhetorical triangle shown earlier! Remember the three golden rules of communication: Keep it brief, keep it simple, and show it with pictures, charts, and graphs.

Figure 9.8 illustrates many of the poster-design principles discussed in this chapter.

This poster is not meant to be a template. It does, however, illustrate several of the principles discussed in the section on poster composition.

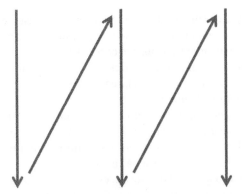

Figure 9.7 The conventional layout of most technical posters.

Include a Brief, Descriptive Title

Names, Credentials, and Affiliations Usually Go Here

Introduction

- Avoid long paragraphs
- Use bullets to reduce word count
- Use a hierarchy of fonts
- Design in 3 or 4 columns
- Text should be readable from 6-8 feet

Methodology

- Remember to use visuals when you can
- Refer to pictures in text (See Figure 1)
- Caption and/or label all pictures

Figure 1: Use Visual Elements When Appropriate (Photos, Charts, & Graphs)

Results

- Use shapes to group related information
- Use color to imply relationships between data
- Label all axes with units
- Edit your charts!
  - Remove any unnecessary lines and information
  - Don't blindly accept charts in the form they are created (See Figures 2 & 3)

Discussion

- This example uses generic IMRaD sections
- Other sections may be necessary
- Even the discussion should be brief, usually in bullet points

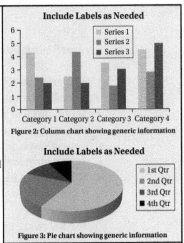

Include Labels as Needed

Figure 2: Column chart showing generic information

Include Labels as Needed

Figure 3: Pie chart showing generic information

Acknowledgements

- Rules regarding plagiarism apply to words, graphics and ideas
- When in doubt, cite!

Figure 9.8 Many of the poster-design principles discussed in this chapter.

Presenting a Poster

Remember that a poster is not always a stand-alone presentation. Although it might have a life beyond the event for which it is composed, it will usually have to be presented in person at a conference or other similar event. The following are some tips for making the most of your poster presentation.

- *Poise*—Don't hide from your audience. Inexperienced poster presenters often try to avoid answering questions about their posters by avoiding eye contact with audience members. This defeats the purpose of composing the poster in the first place! Don't be pushy, but be enthusiastic and friendly. A smile will go a long way toward helping you share your project with a large audience.
- *Message*—Have a two-minute talk prepared that covers the basic elements of your project. When conference goers approach you and your poster, they might have a level of curiosity without having formulated any questions. Your brief talk orients them to your project and introduces them to the elements of your poster.
- *Co-presenters*—If you are presenting a poster as a group, everyone in the group should be conversant on all aspects of the poster, even if the actual work on the project was segmented. Sometimes more than one audience member will engage the group, making it impossible to pass off the presentation to co-presenters.
- *Practice*—Practice presenting the poster to friends and family before you go to the conference. A few practice runs can help you identify the areas where you are weakest in explaining your project, as well as help you find some good phrases to use in your answers.

PowerPoint Presentations

A good presentation renders a service to your audience. A presentation is an opportunity to take the audience from where they are to where you want them to be. The best presentations will do the following:

- Give valuable information the audience probably would not have had otherwise.
- Be in a form that the audience can put into immediate use.
- Motivate and inspire the audience to want to put the presented information into immediate use.

The following tips cover some specific elements of presentations and can help you become a stronger presenter.

The Speaker

Many people would rather do almost anything than speak in front of a group. Being nervous about giving a presentation is normal. It might help you to remember that you aren't expected to be an expert public speaker; you are expected to be a competent engineer. A little nervousness can even endear you to your audience. Just don't let your nervousness take over and ruin your talk. The following list of techniques is intended to help you combat your nervousness.

- Breathing is important; when you get nervous, your respiration rate tends to speed up and your body can go into adrenaline overdrive (a fight or flight mechanism). Controlling your breathing is one way to stay calm.

— Take three deep breaths before you start.
— Practice square breathing, which is inhaling seven counts, holding seven counts, exhaling seven counts, and holding seven counts.
- Visualize yourself successfully presenting your talk.

Practice

If you practice your presentation, you will know how long it takes to complete. If you are nervous, practice will enable you to build in several places in your presentation where you can check yourself and slow down if necessary. The more you practice, the more familiar you become with the material that you are presenting. This familiarity can be important if your information is technical. Showing the audience you are familiar and comfortable with the presented information will help them (and you) gain confidence in your subject. Most likely, you will explain the concepts better as you understand them more fully.

- Have your presentation ready a few days in advance. Being finished usually helps to cut down on your nervousness.
- Rehearse your delivery. When you practice, make it realistic; stand up and actually say aloud the words you intend to say. Many people only practice by scrolling through their slides and going over their lines mentally, perhaps mumbling a few phrases to themselves. This buries your words in the back of your mind. Standing and saying your lines aloud can help you keep the words you wish to say on the tip of your tongue!
- Most people tend to finish their presentations ahead of time because they speak more quickly when nervous.
- Work with your nervousness by tailoring the talk to your strengths. People tend to be most nervous during the initial stages of speaking. One or more of these tips can be used throughout the presentation.
  — Use humor. This makes the atmosphere less formal. Start your presentation with a joke, especially one that pertains to your presentation. However, be careful making jokes if you haven't done everything you were supposed to accomplish. If you were supposed to present ten deliverables, and you only delivered eight, avoid humor. Failure to do what you were supposed to do can make the best joke seriously unfunny!
  — Use props. It gives you something to do with your hands. A show and tell will interest the audience. Be sure the audience can see your prop. If it is too small for people in the back to see, you might be able to pass it around, but sometimes an audience is too large for this method to work smoothly. In fact, if the crowd is too large, sometimes the props can be lost or broken. Weigh the pros and cons carefully before bringing in a fragile or one-of-a-kind prop.
  — Use your nervous habit. Work with that habit to your audience's benefit. If you like jingling change (coins) in your pockets, pull something visual out of your pocket to start your presentation.
  — Use trivia. Focus your audience's attention or teach the audience something interesting that pertains to your talk.

Stage Presence

In the theater, how and where you stand is vital. Because delivering a presentation is essentially a type of theater, these tips can help you make the most of your performance as a speaker:

- Never turn your back to your audience! Not only is it a bit rude, but it also breaks eye contact with your audience and often makes it difficult for your audience to hear you clearly.
- Speak loudly and clearly. If you know that you tend to get nervous and talk too fast when presenting, you can overcompensate a bit by forcing yourself to speak a little slower than your usual pace. It might also help to visualize your words bouncing off the wall at the back of the room. (In smaller rooms, however, be aware that it is possible to be too loud. Nobody likes being shouted at.)
- Plan your choreography if you are presenting in a group. Where will everyone stand? Will you all be on one side of the screen or will you flank it? Will one person advance the slides for all speakers? It's best to answer these questions beforehand. Additionally, your team will look successful if you have a common element in your clothing (for example, everyone wears a white shirt, or has some red in their outfit, such as a tie or scarf).
- Stay out of the projection field. Nobody looks professional or convincing with words on his or her face. If the light from the projector is causing you to squint, then simply step to the side until it doesn't.
- Either stand still or move purposefully. Movements prompted solely by nervousness hurt your stage presence.
  - If you tend to be enthusiastic, use your energy to keep your audience engaged. Make sure, however, that your movements are strong and purposeful. Keep your hands above your elbows if you tend to talk with your hands. Take a few purposeful strides forward as you make strong points. Don't take pointless small steps with stiff arm movements.
  - If you tend to have a more reserved persona, use your poise to project a sense of calm confidence. Stand fairly still, and try to control the pace of your speech. Don't be afraid to move, but if you do move, stay calm, reserved, and controlled in your movements.
- Eye contact is essential. Always avoid talking to the screen instead of to your audience.
  - You can control your audience with your eyes. (If you doubt this, go to a crowded place like a shopping mall, and stare up at the ceiling. It will not take long before you notice everyone around you looking up to see what you are looking at!) Being aware of this control is important because you will keep your audience focused on you, the speaker, by making brief eye contact with audience members throughout the room.
  - In a group presentation, you can do your part when you are not speaking. Any distraction by nonspeakers draws audience eyes away from the speaker. You can help keep audience eyes on the speaker by standing still and looking either at the speaker or at the slides. When you are speaking, make eye contact

with your audience; when you are not speaking, break eye contact with the audience, and look at the speaker or the slides. You will send an additional message of team unity if the non-speakers look in the same place and use the same body language, for example, hands folded in front of you and looking at the speaker.

— If you are explaining data to your audience, it can be difficult to explain specific data on a slide without turning your back to your audience. Using animation to guide your audience's eyes is the best approach. Simply have whatever you want to highlight turn yellow or have a red circle or an arrow appear. However, if you must use a laser pointer or a pointing stick, be sure to point directly to the spot where you want the audience to look, and immediately turn to face your audience again.

Organization

In terms of organizing the information itself, a sample outline of a presentation could be as follows:

- *Introduction*—Who are you, what are your credentials, and why are you presenting?
- *Need*—What is the rationale (need) for your project?
- *Solution*—You have a design or a solution that addresses the need, and here it is. Include major design or solution features and why you did what you did. (Explain how your work specifically addressed the needs.) Be sure to show your final design or explain your solution in its entirety before you begin to address the individual parts.
- *Bottom line*—What are the benefits? How much it will cost?
- *Summary*—Review the main points of the entire presentation. Whatever main concepts you want your audience to remember are the concepts to include on your last slide.

Q & A (Questions and Answers)

Some people like to include a "Questions" slide at the end of the presentation. Rhetorically, this doesn't make sense. Usually, the last point you make will be the most emphatic one. (Think about why your favorite concert performers always play their big hit song last; they want to plant it in your head so you'll go around humming it all week.)

- It is usually better to end on a summary slide or perhaps on an image of your final design than it is to finish on a slide featuring the word "questions," especially because you will usually ask for questions at that time anyway.
- If you have a supervisor who insists on a questions slide, make good use of it. Instead of a blank slide with the word "Questions?" at the top, or one that features cliché clip art, put something useful on the slide, like a computer-aided design and

computer-aided manufacturing (CAD/CAM) image or a schematic. Try to make it something that will prompt thoughtful questions.

- Try to anticipate what questions your audience might ask so that the answers will come to you more easily. Because you are familiar with the contents of your talk, you should know which concepts are difficult to understand or explain. These are areas where you can expect questions. If you can prepare just-in-case slides, slides that are appendices to the main presentation and are only shown if needed during Q & A, then you will look very well prepared to your audience.
- Remember that questions from your audience are a good sign. If your audience has questions for you, that is a good indication that you have held their interest throughout the presentation.
- Never be defensive in answering questions. If someone asks about information you have already presented, resist the urge to respond in an I-already-told-you-so tone. Presentations relay a lot of complex information in a short amount of time; it is understandable that someone might have missed something. Instead, be grateful that the audience member is interested enough in your project to ask.
- If you get caught unprepared for a question, don't dodge or bluff. Admit it when you're stumped. If you made a mistake, own it. Being honest will always make you look better than trying to hide your mistakes. Another technique is to say that you aren't sure of the answer to the question, but that based on what you do know, here's your best educated guess. In this way, people who may be trying to test your knowledge can understand your train of thought.

Equipment

If possible, try to test your slides on the computer you will use to give your presentation. If that is not possible, try to plan for some common problems:

- Always back up your presentation in more than one place and on more than one medium. For example, you can email yourself a copy of your slides as .pptx and .pdf files, and also save them to a portable drive.
- Ensure the proper video projection system is available, up to date, and compatible. You can often bring your own laptop as a backup.
- For overhead projectors, such as ELMOs, if you can read the document when you place it on the projector, then the audience can read it. You might have to focus, but your document will not be backward or upside down.

Slide Design

Many of the best practices for slide design are the same as the best practices for poster design. The biggest difference between the two media is that posters are very limited spaces. PowerPoint slides are free, and although each slide features a finite amount of space, you can use as many slides as are necessary to relay your message.

- If you simply follow the default format that PowerPoint seems to suggest, you will most likely design slides in a topic/subtopic format (see Figure 9.9). This format features a large banner across the top of the slide, and a series of bullet points. When your audience members must remember the individual points, this format can make sense. However, slides of this sort are usually most useful as a teleprompter for the speaker. For the audience, they become tedious very quickly!

Topic/Sub-topic Format Slides

- Sometimes make sense
- Should be limited to short bullets
- Work best when parts of speech are lined up or "parallel"
- Can be more effective when bullets are animated to appear one at a time
- Are often more useful to the speaker as a teleprompter than they are to the audience
- Quickly become tedious to your audience!

Figure 9.9 Topic and subtopic slides.

Topic/Sub-topic Format Slides with Visuals

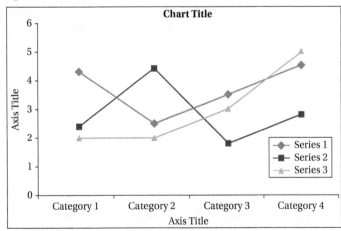

- Work better than words alone
- Often focus more on word than graphics
- Allow for specific details to be read and heard by the audience
- Should follow the same rules as topic/sub-topic slides without visuals
- Also tend to become tedious

Figure 9.10 Topic and subtopic slides with a visual.

- More advanced users of PowerPoint could include some sort of visuals or video clips to supplement this format, such as the one shown in Figure 9.10. The key to a good presentation is often the inclusion of high-quality visual elements; however, a small picture shoved into the corner of a slide does little to overcome the shortcomings of this format.
- Because visuals are the key to a strong presentation, it is a good idea to make the visual elements big, sometimes even taking up the full slide. (Remember Rule 3: If you can show it, show it.) The assertion and evidence slide format (shown in Figure 9.11) makes good use of the entire slide space.
- The assertion/evidence format assumes that most minor points will be made orally. That's fine if a broad understanding on the part of your audience is your goal; however, if the audience needs to understand and retain minor points, animated callout boxes can be added and removed from this format as necessary to give supplemental information, as shown in Figure 9.12. This format allows the audience to see the slide's central idea, the data on which that point is made, and any minor points that the audience might also need to hear.
- It is important to remember that no single slide format works for every occasion. To make better choices, remember the three golden rules of communication: Keep it brief, keep it simple, and show it! Also remember to consider the three questions from the rhetorical triangle: Who is my audience? What is my goal? What does my audience expect from me?

Getting Started

Now that we have talked about everything you can do before a presentation to maximize its effectiveness, we should now think about the presentation itself. The first fifteen to thirty seconds of a presentation are critical; establish eye contact and rapport if you are

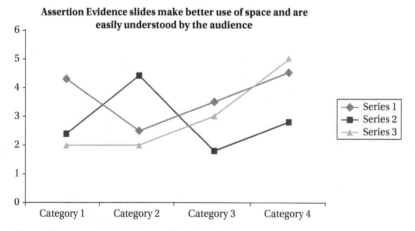

Figure 9.11 Assertion/evidence slide format.

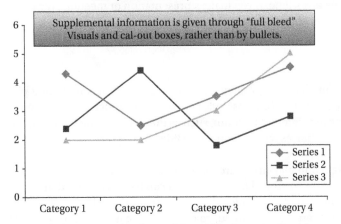

Figure 9.12 Assertion/evidence slides with animated call-out boxes.

going to turn down the lights for slides or a PowerPoint presentation. You can establish a connection with the audience in several ways:

- You can start by saying, "Good morning (or afternoon or evening), my name is [name], and I am going to discuss [subject.]"
- You can establish the level of formality. For example, you can tell the audience that it is okay to ask questions at any time.
- You can establish your tone with humor (telling a joke), or asking a trivia question with a small reward (e.g., a sticker, pen, or other small item) for the winner.
- Interacting with the audience is an excellent way to establish good rapport. After you introduce yourself and the subject of your talk, you can involve the audience by asking a general question. Those interested might raise their hands with an answer. This information could be useful to you (you can learn something about the audience for a better presentation) and can provide you with a good lead into the rest of your talk.
- During your presentation, monitor your audience. Do they look engaged, confused, or bored? If they look engaged, do not change anything that you are doing. If your audience is confused, you could ask them a question, or say, "I know that this information is difficult and some of you look confused. Can I clear up anything for you?" If your audience appears bored, you might reengage your audience with another trivia question or by asking the audience to become active.
- The conclusion of your talk is important. You need to let your audience know that you have finished, for example, "This concludes my presentation. Thank you. Do you have any questions?"

- Stay within time limits! When your talk runs long, you are usually cutting into someone else's time. If you make a presentation or meeting drag on unnecessarily, very few people will appreciate it.

Citations

Most students have no problem remembering the rules regarding plagiarism when writing a paper; however, these rules are often misunderstood or forgotten when communicating in a different medium, such as PowerPoint. The following tips can help you make ethical choices when composing slides:

- Copyright and plagiarism are different issues. Copyright is a legal issue; plagiarism is an ethical issue. Students can use copyrighted material for educational purposes under "fair use" conventions. However, what and how much you can use are limited, so discuss these issues with your instructor for more guidance.
- Many times, you can place a reference on the same slide as the cited material. For example, if you borrow an image from a source, put a brief citation directly below the image. You should, however, have a list of complete references as a part of your slide package. Rhetorically, it makes sense to have it in a set of appendices, rather than to present it as a final slide in your presentation. If you give the acknowledgment on the slide, then you have not plagiarized, yet if anyone wants to see your full references, you will have them ready for the Q & A.
- No single format for citing sources exists in engineering; however, there are several conventional formats that are fairly common. Again, your instructor can give you guidance on this topic.
- Whatever format you use, be consistent! Don't use MLA format on one slide, then Author-Date format on another.
- Remember that anything that is not your original work should be cited, including sounds, images, and ideas.
- When in doubt, cite! It is far better to be safe than sorry.

9.3 **Written Documents**

It is a common misconception among engineering students that once they graduate from college, they won't have to write much. Nothing could be further from the truth. Engineers can expect to write technical reports, proposals, technical memos, résumés, cover letters, and a variety of other documents throughout their careers, and contrary to popular belief, you will not be able to delegate this work to office staff. Understanding what is expected of you in any written rhetorical situation is a vital element to succeeding as an engineer.

Technical Reports

A technical report is a document that explains a technical process, such as an experiment or a design. It is generally broken into three parts:

1. *Front matter*—This section usually includes an abstract, a table of contents, a table of figures, and possibly a list of abbreviations. These elements are like a roadmap letting the audience know where the report will take them, so be careful that the tables and lists are accurate and updated throughout the editing process.
2. *Main body*—This section usually follows a generic format called IMRaD, which is an acronym for Introduction, Methodology, Results, and Discussion.
   — *Introduction*—Details the motivation for the project, scope of the work, and any background information.
   — *Methodology*—Details steps of a procedure or design process, usually making good use of graphics, such as CAD/CAM images, schematics, and photographs.
   — *Results*—Details the findings from the procedures described in the Methodology section. Often, this information is shown in tabular form.
   — *Discussion*—Details the conclusions that can be gleaned from the procedure, as well as any recommendations that follow. The Discussion should discuss larger implications of the work, and should let the reader know what, if any, action is necessary as a result of the project's findings.
3. *End matter*—This section might contain a glossary, an index, and any necessary appendices.

There is a variety of other elements that might be required for a technical report:

- Qualifications of authors
- Budget estimates
- Limitations
- Regulatory requirements
- Glossary of terms
- Specifications
- Research protocols
- Error management
- Literature review
- Technology overview

> Another—perhaps easier—way to think about the main parts of a report is consider an ABC format:
>
> a) Abstract gives a summary of the main points.
> b) Body supplies the supporting details.
> c) Conclusion gives readers what they need to act.

Technical reports differ in style from the kind of writing you might have done in English class. Technical reports are usually written in third-person point of view, meaning they don't use the first-person pronouns *I* and *we*. Technical reports do not directly address the reader using the second-person pronoun *you*. They also avoid flowery prose and complex sentence structure that pays strict attention to poetic elements of rhythm, alliteration, metaphor, and simile. Instead, they follow the three golden rules from the beginning of the chapter:

1. *Keep it brief*—Technical reports avoid unnecessarily complex phrasing. For example, instead of saying "as a result of," a better technical writer will say "because." Instead of redundant phrases like "current status" or "past history," the experienced technical writer will simply say "status" or "history."
2. *Keep it simple*—Instead of long paragraphs, technical writers usually use bulleted lists, which can be read quickly and easily. The best technical writers also avoid being repetitive unless they are trying to emphasize a point. Additionally, headings and subheadings are used to transition from point to point, rather than transitional words or phrases. Headings are hierarchical, with the more important headings in larger, bolder fonts.
3. *Show it*—A technical report usually illustrates its points with pictures, charts, and graphs. These figures should be captioned, numbered, and referred to in the text with phrases like "as shown in Figure 1."

It is important that you fully address the scope of the project in a technical report. If you were asked to cover ten items, it is not sufficient to only cover eight or nine. It is also important that you make sure your information is correct: A well-written wrong answer is useless.

Abstracts and Executive Summaries

One of the unique features of technical reports is the abstract. Abstracts are written for researchers or other specialized readers who need a brief summary of the contents of the full report. A well-written abstract will define the scope of the problem addressed in the paper, will give major findings, and will comment on how these findings relate to theory. An executive summary is essentially the same sort of document, except that its audience is the executives who will often determine the future of your project. For this reason, it will often contain more information related to cost, manpower, and other issues important to one who is overseeing your project.

Abstracts and executive summaries should have the following characteristics.

- Be short, usually a paragraph or two.
- Be written last, after the rest of the report is completed.
- Stand alone, containing all the important points of the report, and none that are not in the report.
- Internally define acronyms and abbreviations.
- Avoid referring to tables, figures, or appendices contained in the main report.

Proposals

Written proposals are documents that describe a need or problem, define a solution, and request funding or other resources to solve the problem. Proposals can be solicited (as when an organization asks for proposals) or unsolicited (as when you send in a proposal and ask for funding or for a project to be implemented). Solicited proposals involve a request for proposal (RFP) or a request for application (RFA), which contain the following guidelines:

- What the proposal should cover
- What sections it should have
- When it should be submitted
- To whom it should be sent
- How it will be evaluated with regard to other proposals

For proposals, you follow the ABC format for report writing, but you concentrate on addressing a need.

- *A: The abstract* gives the summary or big picture for those who will make decisions about your proposal. The abstract usually includes some kind of hook or grabber, which will interest the audience to read further. The abstract will include the following:
  — The purpose of the proposal
  — The reader's main need
  — The main features you offer and related benefits
  — An overview of proposal sections to follow

Commonly Used Proposal Lingo

RFP or rfp: Request for proposal

RFA or rfa: Request for application

PI: Principal investigator; the person responsible for the administration of the project

Co-PI: The persons responsible for the administration of the project. If there are two or more people primarily in charge or with significant duties for a project, then they are co-PIs.

Proposal: The document that you prepare to request funds for a project

Grant: The money that you receive if you've written a successful project

Contract: Most granting agencies require you to sign a contract—a legally binding agreement that you will complete the objectives in the project using the money provided by the agency in a time period specified by the agency.

Program officer: The primary contact person at the agency that is conducting the RFP. Any questions that you might have about your project would be addressed to the program officer. These people can often give you extra information that might not be contained in the RFP.

Proposal reviewer: A person who reads the proposal and critically reviews it. Reviewers check to make sure that the science and engineering are sound, the project is important, the objectives are measurable, the proposed plan of work is excellent, the proposal is well written, and the budget is reasonable.

Resubmission: Grant proposals are hard to get! Usually there are many more potential projects to fund than money to go around. Many federal granting agencies fund only about 10 percent of the projects submitted! This means that some excellent ideas and projects will not receive funding. If your project is not funded, look at the comments from proposal reviewers (these are usually included with a letter telling you that you didn't receive funding). If you do a good job addressing these comments and resubmit your proposal during the next funding cycle, your chances for funding on this resubmission are much better.

- *B: The body* provides the details about your proposal. Your discussion should answer the following questions:
  — What problem do you want to solve or issue do you want to address, and why?
  — What are the technical details of your approach?
  — Who will do the work, and with what?
  — When will it be done, and how long will it take?
  — How much will it cost?

Typical sections in the body portion of the report will be given to you in the RFP, or else you will have to develop them yourself, making sure you address the following:

- A description of the problem or project and its significance
- A proposed solution or approach
- Personnel
- Schedule
- Cost breakdown of funds requested

Give special attention to establishing need in the body. Why should your proposal be chosen over others against whom you are competing? What makes it unique?

- *C: The conclusion* makes your proposal's main benefit explicit and will make the next step clear. This section gives you the opportunity to control the reader's final impression. Be sure to do the following:
  — Emphasize a main benefit or feature of your proposal

— Restate your interest in doing the work
— Indicate what should happen next

Memoranda

A memorandum, or memo for short, is a brief technical communication intended to share information with decision makers. You write a technical memo when you need to document your work in writing. Typical examples include memos to go into a personnel file (congratulations or job improvement details) or those to go into technical files (i.e., engineering projects you are working on). Memos contain introductory information that summarizes the memo's purpose. The memo's conclusion restates the main points and contains any recommendations.

Résumés

A résumé is one of the most important documents you will ever create. It sells you and your qualifications. Generally, there are two types of résumés: skills résumés and experience résumés. A skills résumé is for people who have not yet completed significant work experience. The skills résumé highlights the skills and talents that will benefit a potential employer, even if the applicant has little or no technical work experience. Experience résumés highlight prior work experience related to the job for which a person is applying. In general, most engineering students in the first two years of study and those who graduate without technical work experience will use a skills résumé; engineering students with co-op or internship experience in an engineering context would use an experience résumé.

The following sample and bulleted explanation is intended to assist you with more detailed information on crafting a résumé. A sample evaluation form is included, so you can evaluate your own résumé.

Here are a few important things to remember about résumés:

- A résumé should be crafted for a specific position or job. This approach shows the employer that you care about the position. Your résumé's objective section, if you choose to include one, should reflect some specificity as a result.
- You might need to change your résumé for "scannable format," if you apply for jobs with larger engineering companies. These companies use a computer to scan each résumé submitted to them for keywords. If so, you should consult your career services office or look at the company's recruitment information to determine these keywords and use them on your résumé. The scanner will target you and your résumé as a potential employee more quickly.
- If possible, keep your résumé to one page. Many employers frown on résumés longer than one page.
- Do not include a section entitled "References." Companies assume you have references who can detail your experience and qualifications, so don't waste valuable space on this section.

The format of a résumé is important. Your information should be arranged to be visually pleasing and readable. Use the following strategies in terms of format:

- List dates in reverse chronological order (starting with your most recent information)
- Use at least 11-point font size (preferably 12-point) and professional-looking fonts, such as Times New Roman, Arial, or Calibri
- Use 1-inch margins around the page

Cover Letters

You should always send a cover letter with your résumé. This document introduces you to the company in a more conversational manner. A good cover letter does the following:

- Identifies the position for which you are applying (by number, if so listed)
- Explains why you should be considered for the position (establish a match between the position description and the skills and experience that you possess)
- Highlights specific aspects of your accomplishments, especially those that are noteworthy and will make you stand out with respect to other applicants
- Provides detailed information not in your résumé
- Should not exceed one page

Cover letters contain the following elements:

- Your contact information (address and phone number) and the date
- The full name and contact information of the person at the company or school to whom you must address the letter
- A formal greeting (e.g., Dear Dr. Jekyll or Mr. Hyde)

The opening paragraph should contain an explanation of how you heard about the position and should specifically identify the position itself. For example:

- As you suggested during last week's career fair at <Your University>, I have enclosed a copy of my résumé for your consideration.
- I am submitting my résumé in response to the position description for a consulting engineer (position #EN1234) in the want ads section of <Your local paper> on June 21, 2013.
- You can also state your interest in the position, and briefly highlight your experience, education, or both, that make you well suited to the position.

The central section of the letter will contain one or more paragraphs in which you describe and draw attention to relevant aspects of your education or experience.

- Indicate your knowledge of and enthusiasm regarding the company
- Expand on things especially of interest to the company/school, and things that are important, but that you couldn't detail in your résumé

The concluding paragraph should refer to the résumé for further details, and politely but confidently ask for an interview.

- Include your contact information (phone and email)
- Thank the reader for their consideration
- Include a formal closing, and make sure you sign the letter

The following sample letter contains further information on writing a successful letter.

123 Main Street
Citytown, NY 12345
June 6, 2015

Jane Doe
Manager
ACME Company
321 Maple St.
Citytown, NY 12345

Dear Ms. Doe:
The first paragraph should get right to the point. If you are applying to a job that is listed by a number, include that number in your letter. Your opening paragraph might read something like this: As you suggested during last week's career fair at <Your University>, I have enclosed a copy of my résumé for your consideration. After reading the job description on your company's website, I believe that my skills and my strong work ethic make me a good candidate for the position of field engineer (position #1234).

The second paragraph in any business letter is usually where you go into detail. After having read your to-the-point opening paragraph, the reader can read the middle paragraphs for more specifics. Here is where you can make the items listed on your résumé come to life. Brief narratives about your past experiences and specific skills make a better impression than generic statements about your qualifications. Instead of broadly referring to your leadership skills or work ethic, tell a story where you had to be leader or recount events that demonstrated your high level of commitment. Remember to include details and avoid clichés.

The final paragraph offers an opportunity for follow-up. It might read something like this: I would welcome the opportunity to further discuss this position with you. If you have any questions or need further information from me, please contact me by phone at (555) 555-5555 or by e-mail at jsmith@email.net. I look forward to hearing from you.

Sincerely,

John Smith (Your signature belongs in the space between your closing and your name.)

Enclosure (This lets your reader know that there was an additional document in the envelope: your résumé.)

9.4 **Revising and Editing**

Many people misunderstand what is meant by revision. Some think that every draft of a written document is essentially a restart, and they resent having to do the same task multiple times. The fact is that the process of composing a document is similar to the process of engineering design. In the same way that a draft of an engineering design is not a final design, the first draft of a document should not be a final draft. It should go through the same sort of iterative process of testing, reviewing, and re-designing that engineers regularly employ. Engineers who bring their engineering skills to bear on their communication projects will find far greater success than those who do not.

Whenever you revise a document, it's best to think of big-picture issues first. Although it's always a good idea to catch grammatical errors as you go, it is possible to focus on them too soon. Worrying about commas and periods when you should be figuring out chapters and paragraphs is akin to worrying about what color you will paint a machine before it has been designed.

Note for Group Work

If you are in a group of four, no less than eight eyes should pass over every word in your group report. It is not fair to dump the task of editing onto any one person, especially the poor lead author. All of your names are on the report, so you should all be responsible for its contents. Additionally, you will each have different strengths and weaknesses as writers. This means you will each bring something different to the table when it comes time to revise and edit. This approach will also help to maintain a single voice and tone throughout the paper rather than having a disconnected assemblage of voices, tones, and so on.

Revision

Revision focuses on saying what you want to say and organizing it in the best possible method. Revision is a time for making sure that you have not digressed from your topic, that you are providing the deliverables the task demands, and that you are saying what you need to say in an organized fashion. If you worry about spelling and grammar at this stage, you are likely to spend time correcting sentences that might ultimately need to be cut. You might also be less likely to make changes that need to be made because you consider the sentence "fixed" in terms of grammar.

Make Use of Active Voice

In sentences written in the *passive voice*, the subject is acted upon by someone or something. In those in the *active voice*, the subject of the sentence performs the action expressed by the verb.

- Passive voice: The treatment system *was designed* by the engineer.
- Active voice: The engineer *designed* the treatment system.

Strategy for Finding and Correcting Passive Voice
Passive verbs consist of a "to be" verb (am, is, are, was, were, be, and been) and a main verb.

> Examples: *was conducted, is provided, have been returned*

Step 1: To find verb combinations that might be passive, use the "Edit/Find" feature in Microsoft Word to search for "to be" verbs (am, is, are, was, were, be, & been).

> Passive: An experiment **was** *conducted* by technicians last week.

Step 2: Ask yourself: Is the subject performing the verb, or is the verb happening to the subject? If the subject is performing the verb, your sentence is active and probably doesn't require editing. If the verb is happening to the subject, then the sentence is in passive voice.

> Passive: An experiment *was conducted* by technicians last week.

The experiment isn't doing anything; the technicians are. Therefore, this is passive, and it needs rewording.

Step 3: Reword the sentence.

> Better: Technicians conducted an experiment last week.

Here, the technicians are doing something; therefore, this sentence is in the active voice.

Editing

Once you have gone through several drafts and are sure you are on message and well organized, it is time to turn your attention to details like mechanics, usage, spelling, and punctuation. There are several methods that work well for finding grammatical errors.

- *Method 1*—Read your paper aloud to someone else. Your mouth will catch errors your eyes will not see. When you do this, be sure to have a pencil handy so that when you trip over a phrase you can mark it and see if there was a reason on the page for the mistake.
- *Method 2*—Have someone read your work to you. This will give you an indication of how your words flow through another person's mind. Once again, be sure to mark places where he or she trips over words or has other problems reading aloud. This is often an indication of a problem on the page.
- *Method 3*—Lots of focused editing sessions. Rather than sitting down and trying to catch every error in one sitting, focus on a single issue. For example, read the document once for issues of passive or active voice, and walk away. Come back later to address commas. Another time, hit spellcheck. Lots of short sessions work better than one long grueling one.

9.5 Conclusion

In this chapter, you have learned the basic considerations required for an effective rhetorical strategy. Remember the three questions illustrated by the rhetorical triangle: Who is my audience? What is my purpose? What is expected of me as a communicator? Let these questions guide you through an iterative process regardless of whether you are working on a visual, spoken, written, or multimodal project. Bring your engineering skills to bear on your communication projects, and you will be a successful communicator.

REFERENCES

Alley, M. *The Craft of Scientific Presentations: Critical Steps to Succeed and Critical Errors to Avoid*. Springer: New York, 2003.

Burnett, RE. *Technical Communication*. 3rd ed. Wadsworth: Belmont, CA, 1994.

Cox, MR. *What Every Student Should Know about Preparing Effective Oral Presentations*. Pearson Allyn and Bacon: Boston, 2007.

Markel, MH. *Technical Communication*. 8th ed. Bedford/St Martin's: Boston, 2007.

O'Hair, D, Stewart, RA and Rubenstein, H. *A Speaker's Guidebook: Text and Reference*. 4th ed. Bedford/St. Martin's: Boston, 2001.

Searles, GJ. *Workplace Communications: The Basics*. 5th ed. Longman: Boston, 2011.

Williams, R. *The Non-Designer's Design Book: Design and Typographic Principles for the Visual Novice*. 3rd ed. Peachpit: Berkeley, CA, 2008.

EXERCISES AND ACTIVITIES

9.1   Write a 1-page answer to the following questions.

A.   What color combinations do you see often in your daily life? How do color combinations affect your mood or your thinking about a subject?

B.   Find a photograph from at least twenty years ago. How has society's color palette changed since then?

C.   Look at an advertisement. Did the creator of the ad employ a complementary color scheme? A split complement? Something else entirely?

D.   Look at the clothing you are wearing right now. What does it say to others about who you are as a person? What clothes do you have that might make a different statement?

E.   Think about your most effective teachers. What do they do to make you pay attention to their lessons? Make a list of adjectives to describe the way they present themselves to the class.

F.   What causes you to lose faith in someone's message? What sorts of behavior make you roll your eyes in contempt at a speaker? How might you avoid these pitfalls as a speaker?

G.   Think about a PowerPoint presentation you either must give soon or recently have given. How would you present that information if you were not able to use any words on your slides?

H.   It is easy to see how a technical report or a proposal is different from a story or a poem. How are they similar?

I.   What grammatical rule or principle do you routinely get wrong? How can you learn that rule once and for all?

J.   How do you use writing in your everyday life, outside of school and work (grocery lists, texting, diary entries, etc.)? How can you make similar writing acts useful to you as an engineer?

9.2   Interview one of your professors or a practicing engineer to learn more about speaking skills. You might ask them some or all of the following questions:

- Do you ever get nervous before a public speaking engagement? If so, what do you do to relax? What advice do you have for me to relax if I get nervous?
- How do you prepare for an oral presentation?
- Do you employ any methods for engaging the audience at the beginning of a talk?
- How do you keep your audience interested throughout your talk?
- Do you have any advice for me as a public speaker?

9.3   Go to a public speaking engagement at your university with a speaker of regional or national prominence. Critique this person's speaking style with respect to the guidelines for oral presentations presented in this chapter. What did the speaker do particularly well? What could he or she have improved on? What things did the speaker do that you would like to integrate into your own speaking repertoire?

9.4 Build a paper airplane or other origami figure and write out the directions for building it as you go. Then give your directions to a friend or colleague and ask them to build the same paper airplane or origami figure. Watch them as they build and compare their final product to yours. What did they do correctly? What did you do well to enable their success? Where did they go wrong? What might you have done differently to help them avoid their mistakes?

9.5 Share a recent slide or poster presentation with a friend or colleague via the Internet. How does using Internet conferencing software (e.g., Skype, GoToMeeting, Connect, etc.) change the way you present your information? What parts of the presentation convert easily to a virtual environment? What new challenges does the virtual environment raise?

9.6 Choose an organization that you are interested in working with for a specific purpose, for example, a community project, summer internship, co-op, or future employer. Write a résumé and cover letter to this organization.

CHAPTER 10

Ethics and Engineering

10.1 Introduction

In addition to technical expertise and professionalism, engineers are also expected by society and by their profession to maintain high standards of ethical conduct in their professional lives. This chapter will cover the most important issues of ethics insofar as they are relevant to careers in the various engineering disciplines.

10.2 The Nature of Ethics

Generally, ethics is concerned with standards, rules, or guidelines for moral or socially approved conduct such as being honest or trustworthy, or acting in the best interest of a society.

Not all standards or values are ethical standards, though. For example, personal preferences and values such as individual choices of food or clothing aren't important enough to qualify as ethical values or choices. Generally speaking, ethical standards apply only to conduct which could have some significant effect on the lives of people in general.

Thus for an engineer, using a substandard grade of steel in the construction of a bridge would definitely violate ethical standards because of the potential for safety hazards for people in general. But on the other hand, using an inferior brand of ketchup on one's fries does not violate any ethical standards, since at worst it would only violate personal, non-ethical standards of what tastes good to oneself.

Ethics and ethical standards should also be distinguished from matters of legality and legal standards. Roughly speaking, the distinction is that legal standards are defined in legal documents by some properly appointed legal body, and those documents and legal experts determine what the law is and who should obey it.

In ethics, by contrast, it is instead assumed that ethical standards exist independently of any particular group of experts (being accessible to all through the exercise of their own thinking capacities), and that a codification or written-down form of the standards merely describes or summarizes what those pre-existing ethical standards are, rather than (as in the case of the law) defining their very nature.

For example, if there were a legal requirement that any load-bearing beam in a bridge must be able to bear five times the average real-world stress on the beam, then it is a simple matter of calculation to determine whether a beam meets the legal standard or not. However, an ethical standard for what would be an adequate safety factor in such a case is a very different matter, and it cannot be settled by calculation or appeal to a rule book, since an ethically concerned engineer could convincingly argue that conventional or routine safety factors are really ethically inadequate and unacceptable. Such a person cannot be answered merely by appeal to some rule book or to current legal standards for such matters.

Another vital point about ethics is that its standards are always more important than any other standards. Thus in the contrast between personal and ethical standards discussed above, if one's personal standards conflict with ethical standards, then one must suppress one's personal standards to resolve the conflict in favor of the ethical standards. For example, if one personally values finding the highest-paying job, but it turns out that the job in question involves some ethically wrong conduct, then it is one's ethical duty not to accept that job, but to take a lower-paying, more ethical job instead.

It is generally held that even legal standards must give way to ethical standards in the case of conflicts. For example, suppose that a law governing motor vehicles did not currently require a vehicle recall in the case when engineering defects were found in its construction. Nevertheless, if a good case could be made that this was an ethically unacceptable situation, then the law would have to be changed to conform to the ethical standard. So here too, other kinds of values or standards must be suppressed or adjusted in the case of a conflict with ethical standards, which are always of overriding importance.

10.3   **The Nature of Engineering Ethics**

Much of engineering ethics is an applied or more specific form of what could be called "general ethics," that is, ethical standards which apply to any human activity or occupation. For example, ethical duties of honesty, fair dealing with other people, obeying the relevant laws of one's country or state, and so on apply in any situation. Thus some of the ethical obligations of engineers are of this general kind also found in many other occupations and activities.

Other kinds of standards found in engineering ethics are more specific but not unique to engineers, such as the principles governing the ethical activities of any professionals in their contacts with clients or customers. These too are an important part of engineering ethics, and they show up in the large overlap in the contents of many codes of ethics derived from many different professions (engineering, law, medicine, academia, etc.).

Then finally there are some standards applying primarily to the profession of engineering, such as those dealing with the proper, ethical manner of approval of designs or plans by managers which require them to have professional engineering qualifications and expertise. These standards are what make engineering ethics distinctive relative to other kinds of applied ethics, but they are only one small part of the whole of engineering ethics, and so should not be overemphasized.

Now to the question of how ethical standards for engineers differ from other standards of value which engineers might use. This brief discussion will parallel that given above for ethics. The aim is to get clear about which of the choices an engineer might make that are relevant to engineering ethics, as opposed to other choices which are instead based on other kinds of values or standards.

As with ethics in general, purely personal choices or values are not relevant to engineering ethics. For example, an engineer who is submitting a bid to a potential client on an engineering project might personally decide to factor in a somewhat higher rate of profit on the project for him/herself than is usual in such projects.

However, as long as there is an open bidding process (where others are free to submit possibly lower bids), and as long as the personal pricing would not affect the quality of the work which would be done, then the pricing level remains a purely personal or economic decision with no ethical implications.

On the other hand, if an engineer were to decide to maximize his/her profit in another way—by submitting a low bid, but then secretly drastically compromising on the quality of materials used so as to achieve maximum personal profit—that would be a violation of engineering ethics standards, which generally require the use of high-quality materials and construction methods acceptable to all parties signing the initial project agreement. This would be a case in which a conflict between personal standards (maximum profit) and engineering ethical standards (honoring the contract and using high-quality materials and methods) must be resolved in favor of supporting the relevant engineering ethics standards in this case.

These contrasts between personal versus ethical values apply also to the related contrast of corporate or economic values versus ethical values. Thus the examples just given of legitimate versus ethically illegitimate maximizing of profits would apply just as well if the motivation were the interest of some corporation or economic group—it would be equally wrong for a corporation or engineering firm to seek profit at the risk of lower engineering quality of products or services. Thus it is the duty of an engineer to uphold engineering ethics standards even if (to take an extreme case) his/her job is at risk, such as when doing the ethically right action would conflict with the non-ethical interests of a boss or corporate structure.

Another area of close harmony between different kinds of values in engineering ethics is found in the case of technical or scientific values. Any scientific community dealing with pure sciences or applied sciences such as engineering will have technical values concerning the proper and correct ways to construct theories, carry out calculations, make statistically satisfactory estimates where exact results are not possible, and so on. These themselves are not ethical values (good science is not the same as good ethics). But nevertheless engineers do have an ethical obligation to use good scientific methods at all times, and so here too (as with the law) there is a close correlation between practicing good science and being an ethical engineer.

One other topic should be mentioned here. Naturally, most of engineering ethics concerns an engineer's behavior while engaged in professional engineering activities. But one should be aware that engineering ethics codes also generally include prohibitions on unethical behavior while off the job as well, if those activities would affect public perceptions of one's professional integrity or status. This would include activities such as gambling which might tend to bring the profession of engineering into disrepute, deceptive or other inappropriate forms of advertising of services, or any other activities suggesting a lack of integrity or trustworthiness in an engineer.

EXAMPLE 10.1

Engineers should follow their professional code of ethics because:

A) It helps them avoid legal problems, such as getting sued
B) It provides a clear definition of what the public has a right to expect from responsible engineers
C) It raises the image of the profession and hence gets engineers more pay
D) The public will trust engineers more once they know engineers have a code of ethics

Solution: Choice A: Avoiding legal problems is generally a good thing, but it's not the most relevant reason why the code should be followed (and in some cases, strictly following the code might make it more likely that you'd get sued, e.g., by a disgruntled contractor if you refuse to certify shoddy workmanship).

Choice C: Raising the image of the profession is a good thing, but not for the reason of getting more pay; that would be to act from self-interest rather than ethical motivations.

Choice D: Increased public trust is generally a good thing, but the mere knowledge that there is such a code means little unless engineers actually follow it, and for the right reasons.

Choice B: The best and most relevant answer. The code of ethics is designed to promote the public welfare, and hence the public has a right to expect that responsible engineers will follow each of its provisions, as clearly defined by the code.

EXAMPLE 10.2

Engineers should act ethically because:

A) If they don't, they risk getting demoted or fired
B) The boss wants them to
C) It feels good
D) That's the way responsible engineers behave

Solution: As with Example 10.1, some of the wrong answers here do have some connection with ethics, so as before the task is to pick the best and most relevant answer.

Choice A: Even though ethical behavior may usually improve one's job security, it's not the most relevant reason for being ethical.

Choice B: Doing what the boss wants is a self-interested reason for acting, not an ethical reason.

Choice C: Even though, hopefully, being ethical will make engineers feel good, it's not the main reason why they should be ethical.

Choice D: The right answer. There's a very close connection between acting in a responsible manner and acting ethically.

EXAMPLE 10.3

The first and foremost obligation of registered professional engineers is to:

A) The public welfare
B) Their employer
C) The government
D) The engineering profession

Solution: As before, the best and most relevant answer should be chosen.

Choice B: Engineers do have ethical obligations to their employer, but it's not their foremost or primary obligation.

Choice C: As with the employer, engineers do have some obligations to the government, but again, not the first or primary one.

Choice D: Obligations to the engineering profession are probably even more important than those to employers or the government, but still they are not the foremost obligation.

Choice A: Correct. The foremost, primary obligation of engineers is to the public welfare.

EXAMPLE 10.4

Registered professional engineers should undertake services for clients only when:

A) They really need the fees
B) Their own bid is the lowest one
C) They are fully technically competent to carry out the services
D) Carrying out the services wouldn't involve excessive time or effort

Solution: Choice A: Personal financial need is a matter of self-interest, not of ethics.

Choice B: The competitive status of a bid is a matter of economics, not ethics.

Choice D: The amount of time involved in carrying out services is also a matter of economics and self-interest, rather than of ethics.

Choice C: Correct. Ethical engineers will not undertake to provide services for clients unless they are fully technically competent to carry them out.

10.4 The Issues and Topics of Engineering Ethics

Now we will examine the basic issues, concepts, and special topics of engineering ethics. Recall that most of these are also included in or assumed by the codes of ethics provided by various professional engineering societies. Hence as noted in the Introduction, we will use the most relevant one—the NCEES Model Rules of Professional Conduct—as the basis for the discussion. Its study will provide a good understanding of the typical actual ethical requirements imposed by professional engineering societies on their members.

The Preamble

The Preamble (the Introductory section) to the Rules is important in that it describes their purpose, which is to safeguard life, health and property, to promote the public welfare, and to maintain a high standard of integrity and practice among engineers. Where it is helpful, we will slightly paraphrase the statements taken from the code so as to clarify their meaning.

This part of the code makes explicit its connection with general ethical concerns which all ethical persons should adhere to. We will examine each of the items covered in turn in this statement of the purpose of the code.

1. First, normative ethical theories universally agree that it is ethically wrong to cause harm to people.

This is why the Preamble mentions the purpose of "safeguarding" life, health, and property. Thus ethical persons in general, including engineers, should avoid doing anything which would damage or adversely affect other people. More positively, one should also take measures which will safeguard or preserve people from possible future harm.

For engineers, the more positive measures would include such things as building devices with extra "fail-safe" features included which make harmful consequences of their use as unlikely as possible. Overall, then, ethical persons take great care not to cause harm to others, and they take whatever extra steps are necessary to minimize risks of potential harm to others as well.

2. The second purpose of the code is to " . . . promote the public welfare."

Here too, standard normative ethical theories fully support this rule. It states a duty or obligation not simply to act in a harmless or safe way (as previously discussed), but in addition to take active steps so that one's professional activities will result in definite benefits and improved conditions for the general public.

For example, as an engineer planning a new highway this rule would require one not only to plan and build it in a safe manner (as previously discussed), but also to do such things as choose the shortest feasible route between its endpoints, or to choose that route which would permit the most efficient road-construction techniques, hence maximizing the utility of the highway to the general public and minimizing its cost to them as well.

3. The third and last purpose of the code, according to the Preamble, is " . . . to maintain a high standard of integrity and practice among engineers."

Thus the code also has the purpose of ensuring that engineers will continue to be honest and trustworthy, and to maintain high standards of professional conduct and scientific expertise in their work.

Note that this Preamble makes it explicit that the reason why engineers should adhere to the code isn't simply because it is ethical to do so, or because their professional organization tells them to do it. Instead, it emphasizes that there are vital practical benefits to society which can only be achieved by engineers committing themselves to rigorously follow the code at all times.

Other Issues

Other issues covered in the Preamble are as follows:

> *"Engineering registration is a privilege and not a right. This privilege demands that engineers responsibly represent themselves before the public in a truthful and objective manner."*

A privilege is a socially earned license to do something which is granted by society only under certain conditions. For example, a driving license requires the passing of a driving test and other requirements. Similarly, engineering registration must be earned by recipients (they do not automatically have a right to that status) and the granting of it requires them in return to adhere to ethical standards such as the stated one.

> *"Engineers must compete fairly with others and avoid all conflicts of interest while faithfully serving the legitimate needs and interests of their employers and clients."*

This statement sums up a range of ethical requirements which are more fully covered in the body of the code, so the various issues which it raises will be discussed along with those requirements below.

The Engineer's Obligation to Society

The first group of rules in the code address **the engineer's obligation to society**. As be- fore, we will slightly paraphrase the statements taken from the code so as to clarify their meaning:

1. While performing services, the engineer's foremost responsibility is to the public welfare.

This rule is also featured in a related form in the Preamble, as a duty to promote the public welfare. The idea here of responsibility to the public welfare includes the idea of safeguarding the public from harm, as more fully spelled out in the next rule:

2. Engineers shall approve only those designs that safeguard the life, health, welfare and property of the public while conforming to accepted engineering standards.

These two rules together imply a much broader context of responsibility for engineers than those arising from any one task or project. Designs and materials that seem perfectly adequate and ethically acceptable within the bounds of a given project may nevertheless be unacceptable because of wider issues about the public interest.

For example, until recently a refrigeration engineer could have specified Freon (a chlorinated fluorocarbon or CFC product) as the prime refrigerating agent for use in a product, and defended it as an efficient, inexpensive refrigerant with no risks to the purchaser of the appliance. However, we now know that there are significant risks to the public at large from such chemicals because of the long-term damage to the environment they cause when they leak out, perhaps many years after the useful life of the product is over. Rules 1 and 2 tell engineers that they must always keep such wider, possibly longer-term issues in mind on every project they work on.

3. If an engineer's professional judgment is overruled resulting in danger to the life, health, welfare or property of the public, the engineer shall notify his/her employer or client and any authority that may be appropriate.

An important rule, which may place the engineer in a difficult position if his/her employer or client is among those contributing to the problem. But the engineer's duty in such cases is clear: ". . . any authority that may be appropriate" must be notified, even if the employer/client tries to prevent it. (Cases of this kind are popularly referred to as "whistle blowing.")

4. Engineers shall be objective and truthful in professional reports, statements, or testimonies and provide all pertinent supporting information relating to such reports, statements, or testimonies.
5. Engineers shall not express a professional opinion publicly unless it is based upon knowledge of the facts and a competent evaluation of the subject matter.

Rules 4 and 5 together implement the general ethical requirement that one ought to tell the truth in the specific context of an engineer's professional duties. Note that the duty as mentioned in Rule 4 is not simply to be truthful in what one says, but also to be forthcoming about all pertinent or relevant information in reports, etc.

Rule 4 also mentions being "objective," which adds the element of being unbiased and basing one's beliefs and reports only on objective, verifiable matters of fact or theory.

Rule 5 enlarges on the idea of being objective in one's reports—others should be able to rely upon one's professional opinion, and they can do this only if one knows all of the relevant facts and is completely competent to evaluate the matter being dealt with.

6. Engineers shall not express a professional opinion on subject matters for which they are motivated or paid, unless they explicitly identify the parties on whose behalf they are expressing the opinion, and reveal the interest the parties have in the matters.

Rule 6 expresses what is sometimes called the duty of full disclosure. Even if one honestly seeks to be truthful and objective (as in Rules 4 and 5), still doubts might be raised about one's motivation or objectivity unless one reveals on whose behalf one is expressing an opinion, and the interests that such persons have in the case. This rule is also related to the issue of conflicts of interest (see Rules 6 and 8 in the second section, below).

7. Engineers shall not associate in business ventures with nor permit their names or their firms' names to be used by any person or firm which is engaging in dishonest, fraudulent, or illegal business practice.

This might be called the "clean hands" rule (shake hands only with those whose hands are as ethically clean as your own). It isn't sufficient to be completely ethical in one's own (or one's own company's or firm's) practices; one must also ensure that others do not profit from one's own good name if their own activities are unethical in some way. This Rule 7 is clearly related to Rule 1 concerning the public welfare—one must promote this in one's external dealings just as much as in one's own activities.

8. Engineers who have knowledge of a possible violation of any of the rules listed in this and the following two parts shall provide pertinent information and assist the state board in reaching a final determination of the possible violation.

Rule 8 generalizes Rule 3 (a duty of disclosure when one's professional judgment is overruled) to a duty of disclosure in the case of any of these rules, when one has knowledge of possible violations of them. In terms of the public welfare, it is very important that each profession regulates itself in this way, so as to minimize or eliminate future infringements of its rules. Strict adherence to this rule also will lead to wider appreciation and respect for the profession of engineering because of its willingness to "clean its own house" in this way.

The Engineer's Obligation to Employers and Clients

The second group of rules in the code address **the engineer's obligation to employers and clients**.

1. Engineers shall not undertake technical assignments for which they are not qualified by education or experience.
2. Engineers shall approve or seal only those plans or designs that deal with subjects in which they are competent and which have been prepared under their direct control and supervision.

Rules 1 and 2 require an engineer to be professionally competent, both in undertaking technical assignments and in approving plans or designs. Rule 2 in fact requires a double kind of knowledge, both technical competence in the matters to be approved and that one has had direct control and supervision over their preparation. Only thus can one be sure that one's approval is legitimate and warranted.

3. Engineers may coordinate an entire project provided that each design component is signed or sealed by the engineer responsible for that design component.

Rule 3 in effect invokes Rule 2. As long as each component of a project is satisfactorily approved as per Rule 2, then it is permissible for an engineer to coordinate an entire project. This also underlines the importance of Rule 2, as project managers have to rely heavily on the validity of the approvals for each prior part of a project.

4. Engineers shall not reveal professional information without the prior consent of the employer or client except as authorized or required by law.

This confidentiality requirement is the other side of the coin of duties to tell the truth (as in Rules 4 and 5 of the first Obligations to Society section). Just as one must not lie or misinform, so also must one restrict to whom one reveals professionally relevant information. A professional engineer must be trustworthy as well as honest, and being trusted not to reveal confidential information is an important kind of trust.

Confidentiality is a central factor in assuring employers and clients that one's professional services for them are indeed for them alone, and that they can rely upon one's discretion in not revealing to others any private information without their full consent. This rule is also related to Rules 5 through 8 below, in that any revealing of information to others would probably create conflicts of interest or other "serving more than one master" problems of those kinds.

5. Engineers shall not solicit or accept valuable considerations, financial or otherwise, directly or indirectly, from contractors, their agents, or other parties while performing work for employers or clients.

This is the first of four rules dealing with problems of "conflicts of interest." These are cases where one has some primary professional interest or group of interests—generally, to carry out some project for an employer or client—but where other factors might enter into the picture which would activate other, non-professional interests one also has which would then conflict with the professional interests.

In the case of Rule 5, the concern is that soliciting or accepting such things as gifts, hospitality, or suggestions of possible future job offers for oneself would activate non-job-related, personal interests of yours (for additional pay, career advancement, etc.) which would then be in conflict with one's primary professional interests and duties concerning a current project.

Special attention should be paid to this and the other conflict-of-interest rules, and they should be strictly observed. People are sometimes tempted to think that breaking these rules is ethically harmless, on the grounds that if one has a strong enough character, then one will not actually be professionally influenced in a detrimental way by gifts, etc., and hence that accepting such inducements cannot do any harm.

However, it is important to realize that even the *appearance* of a conflict of interest (however careful one is to avoid actually undermining one's professional interests) can create serious ethical problems. One basic concern about this is the potential loss of trust it could cause in an employer or client. Just as clients need to know that the engineer will keep their information confidential (as in Rule 4), so also they need to know that the engineer is single-mindedly working with only their interests at heart.

Any doubts raised because of the appearance of a conflict of interest could be very damaging to the client/engineer professional relationship. This is why it is necessary for the professional engineer to avoid doing anything which would create conflicts of interest or even the appearance of them.

Further rules are necessary to deal with conflicts of interest, because unfortunately in some situations the appearance or possibility of conflicts of interest may be virtually unavoidable, no matter how ethically careful everyone is. However, fortunately at the same time there is a powerful method available for minimizing any ethically bad effects of such situations. This is the method of full disclosure of potential conflicts to all interested parties, and it is addressed in the following two rules:

6. Engineers shall disclose to their employers or clients potential conflicts of interest or any other circumstances that could influence or appear to influence their professional judgment or the quality of their service.
7. An engineer shall not accept financial or other compensation from more than one party for services rendered on one project unless the details are fully disclosed and agreed to by all parties concerned.

Both these rules address issues of full disclosure, or keeping all the relevant parties fully informed as to areas of potential conflict or potentially undue external influences. The basic idea behind full disclosure is that it can maintain trust and confidence between all parties in several important ways, as detailed below.

First, if engineer A informs other party B about a potential conflict or influence, then A has been honest with B about that matter, hence maintaining or reinforcing B's trust in A. Furthermore, if B is not further concerned about the matter once he/she knows about it, then the potential problem (namely, the apparent conflict or potentially bad influence on A's professional conduct) has been completely defused.

Suppose on the other hand that B is initially concerned about the issue even after it was honestly revealed to B by A. Even so, the problem is already lessened: at least A has honestly revealed the area of concern. Things would be much worse if B later discovered the problem for him/herself, in a case when A had not fully disclosed it—that would be very destructive of trust between A and B.

Furthermore, now that B knows about the area of concern, and knows that A is fully cooperating with him/her in disclosing the potential problem, both of them can proceed to work out mutually acceptable ways of minimizing or disposing of the problem to their joint satisfaction. Thus even if the full disclosure does lead to an initial problem which needs to be resolved, nevertheless it is one which does not break down the trust between A and B. In fact it may even reinforce trust, in that A's willingness to fully disclose a potential conflict/influence and negotiate with B about it is good evidence for B of A's professional honesty and sincerity.

8. To avoid conflicts of interest, engineers shall not solicit or accept a professional contract from a governmental body on which a principal or officer of their firm serves as a member. An engineer who is a principal or employee of a private firm

and who serves as a member of a governmental body shall not participate in decisions relating to the professional services solicited or provided by the firm to the governmental body.

Rule 8 deals with a special case of potential conflicts of interest, namely, when one of the interested parties is a governmental body. In such a case, a somewhat stricter rule is required than for the more usual cases involving only non-governmental agencies.

In non-governmental cases, it is ethically sufficient to fully disclose potential conflicts of interests to all parties, and then to negotiate with the other parties as to how to deal with the potential conflicts. For example, if one's engineering firm has an official who was also on the board of directors of a bank, it would be ethically acceptable to accept a professional contract from that bank, as long as all parties were fully informed about the official's joint appointment prior to the agreement, and also as long as they could come to agree that the joint appointment was not an impediment to their signing a contract.

Returning to the governmental case, the reason why the stricter Rule 8 is required when one of the parties is a governmental body is as follows: In the case of agreements among private persons or businesses, they themselves are the only parties having a legitimate interest in the negotiations, and hence whatever they freely decide among themselves (of course assuming that no other ethical rules or laws are being broken) is acceptable.

On the other hand, in the case when a governmental body is involved, there is another interested party which is not directly represented in negotiations, namely the electorate or citizens of the jurisdiction covered by that governmental body. The governmental body must act only in ways which fully respect the interests and concerns of the electorate. In such a case, it is impossible to ensure that full disclosure to all of the citizens of the electorate of the potential conflicts of interest would be made in a case such as that envisaged in Rule 8, and hence there is a need for the stricter rule which completely prohibits conflicts of interest of the kinds defined in Rule 8. Only then can public trust both in the engineering profession and in governmental bodies be preserved.

An Engineer's Obligations to Other Engineers

The third group of rules in the code primarily address **an engineer's obligations to other engineers**. These rules can be considered as more precise specifications of various rules already introduced, when applied to the specific context of obligations to other engineering professionals. The first rule also covers obligations to potential employers (whether they are engineers or not) when one is seeking employment.

1. Engineers shall not misrepresent or permit misrepresentation of their or any of their associates' academic or professional qualifications. They shall not misrepresent their level of responsibility nor the complexity of prior assignments. Pertinent facts relating to employers, employees, associates, joint ventures or past accomplishment shall not be misrepresented when soliciting employment or business.

This Rule 1 is an application or more specific form of Rules 4 and 5 of the first section (concerning an engineer's obligation to society), which require objectivity and truthfulness in all professional reports, statements, and opinions.

Rule 1 requires one not to misrepresent one's own qualifications, or those of associates, which is a very important kind of truthfulness and objectivity. The rule further specifies that this duty not to misrepresent applies also to issues of prior levels of responsibility and to the complexity of previous assignments as well. The rule concludes with a statement about any and all pertinent facts in one's previous history, and it requires that they too should not be misrepresented.

It might be thought that this Rule 1 is unnecessary, since it simply applies some general ethical principles already in the code to special cases and circumstances centering around issues of employment and qualifications. However, as with issues of conflicts of interest, there are some special temptations in these areas which are best addressed by explicitly spelling out what is ethically required for dealing with such situations.

A typical temptation in this area might go something like this: Even a well-meaning, otherwise generally honest engineer might be tempted to make his/her qualifications seem more impressive to a potential employer. One might rationalize that one really would be the best person to do the required job, and that therefore one is doing the employer a favor by making one's qualifications seem more impressive and thereby inducing them to hire oneself, rather than some less-qualified person whose paper qualifications might misleadingly look as good as one's own.

Clearly there could be many variants on this manner of thinking. But all would involve rationalization (the inventing of dubious reasons for what one wants to believe anyway) and wishful thinking, rather than the objectivity and true rationality and truthfulness which are required of professional engineers. So Rule 1 serves a useful function in explicitly stating requirements which some might otherwise be tempted to ignore or overlook.

2. Engineers shall not directly or indirectly give, solicit, or receive any gift or commission, or other valuable consideration, in order to obtain work, and shall not make a contribution to any political body with the intent of influencing the award of a contract by a governmental body.

Rule 2 continues the prohibitions against conflicts of interest which were found in Rules 5 through 8 in the previous section (rules covering the engineer's obligations to employers and clients). Those previous rules mainly covered cases where an engineer was already employed, while Rule 2 here specifically applies to attempts to obtain future work, including the award of a contract by a governmental body.

This Rule 2 also emphasizes that it is just as wrong to attempt to unduly influence someone else (a potential employer, for instance) as it would be to allow others to unduly influence oneself. Thus the rule underlines that it is just as ethically unacceptable to try to cause conflicts of interest in others as it is to allow oneself to be enmeshed in improper conflicts of interest.

It should be noted that the second part of Rule 2, ". . . and shall not make a contribution to any political body with the intent of influencing the award of a contract by a governmental body," specifically mentions the intent of the person making the contribution. Contributions are not prohibited, only contributions with the wrong intent.

This part of the rule could be difficult to apply or enforce in practice, because it may be very hard to establish what an engineer's actual intent was in making a political contribution. Also, the freedom to make political contributions to organizations of one's own choice is generally viewed as an ethical right which should be limited as little as possible. So this part of the rule very much depends on and appeals to the ethical conscience of the individual engineer, who must judge his/her own intentions in such cases and avoid such contributions when their own intent would be self-interested in the manner prohibited by the rule.

Note also that the first part of Rule 2, "Engineers shall not directly or indirectly give, solicit, or receive any gift or commission, or other valuable consideration, in order to obtain work . . . ," also mentions the reason or intention behind giving or receiving gifts, etc., in the phrase "in order to obtain work." However, in practice it is much easier to judge when gift-giving is ethically unacceptable than when political contributions are unacceptable, since there are more behavioral and social tests for suspicious inducements to obtain work than there are for suspicious political support. So it is much easier to police and regulate infringements of this first part of Rule 2 than it is for the second part.

3. Engineers shall not attempt to injure, maliciously or falsely, directly or indirectly, the professional reputations, prospects, practice or employment of other engineers, nor indiscriminately criticize the work of other engineers.

Rule 3 as it stands is somewhat unclear, so some discussion is required to bring out its ethically legitimate core.

First, how should the subordinate phrases "maliciously or falsely" and "directly or indirectly" be interpreted? On one possible interpretation, Rule 3 says outright that engineers should never attempt to injure in any way or for any reason the professional reputations of other engineers. On this interpretation those phrases just give examples of possible modes of injury which are prohibited, leaving unmentioned any other possible modes of injury which are nevertheless also assumed to be prohibited.

However, another interpretation is possible, according to which it is only certain kinds of injury which are prohibited by Rule 3, namely those spelled out by those same phrases interpreted so that the "directly or indirectly" part modifies the "maliciously or falsely" part. On this interpretation, Rule 3 prohibits only malicious or false attempts (whether carried out directly or indirectly) to injure the reputations of other engineers.

This second interpretation would ethically permit attempts to injure the reputations of other engineers, as long as the attempts were carried out in a non-malicious and honest, truthful way (and presumably with the public welfare in mind as well).

Some support for this second interpretation can be derived from the final clause of Rule 3, "nor indiscriminately criticize the work of other engineers." Clearly, in this case it is not all criticism of the work of other engineers which is being prohibited, but only indiscriminate criticism. Thus the last section of Rule 3 outlaws criticism which is over-emotional, biased, not well reasoned or factually inaccurate, and so on, but it does not prohibit well-reasoned, careful, accurate, factually-based criticisms of other engineers.

Thus in support of the second interpretation, it might be said that there too, it is not all attempts to injure reputations, which are being prohibited, but only those which are malicious, false, indiscriminate, or otherwise highly ethically questionable in the methods they employ.

However, there is still something to be said in favor of the first interpretation as well (which, it will be recalled, involves a blanket condemnation of all attempts to injure reputations). Those defending it might do so as follows: It is arguable on general ethical grounds that any attempt to injure someone's reputation must be viewed as going too far and therefore becoming unethical. Even if one is convinced that someone else's work is shoddy, dishonest, and so on, at most (it could be argued) one has a duty to point out the problems and shortcomings in their work. It is a big leap from criticizing an engineer's actions, on the one hand, to condemning the engineer in a way designed to injure his/her reputation, on the other hand.

Thus (on this line of thinking) one should criticize an engineer only out of a disinterested desire to let the truth be known by all, not with the aim of injuring someone's reputation—it is for others to judge whether the truth of what one has pointed out will diminish the reputation and so on of the engineer in question.

Fortunately, it is not necessary to definitively decide between these different interpretations of Rule 3. What is important is to become sensitive to the ethical issues involved in each interpretation. And for the purposes of conforming to Rule 3, both sides can agree that if criticism of other engineers ever becomes necessary, it should be done in a very cautious and objective manner, and with all due respect for the professional status of the person being criticized.

EXAMPLE 10.5

With respect to the Moral Rules of Professional Conduct for engineers:

A) The rules are a bad thing because they encourage engineers to spy on and betray their colleagues
B) The rules are a useful legal defense in court, when engineers can demonstrate that they obeyed the rules
C) The rules enhance the image of the profession and hence its economic benefits to its members
D) The rules are important in providing a summary of what the public has a right to expect from responsible engineers

Solution: Remember here that the best/most relevant answer is to be chosen, because each answer has some truth to it, but only one has the most truth.

Choice A: It is true that the rules require those who have knowledge of violations of the rules to report such cases to the relevant State Board. And this could involve one in collecting more information on the possible infringements, and hence exposing those involved to the scrutiny of the State Board. However, the generally negative connotations of "spying" and "betraying colleagues" do not make the rules bad—the activities in question are a necessary part of responsible reporting of possible violations to the authorities, and hence are morally fully justified.

Choice B: It is true that proof in court that one has followed the rules may be a useful legal defense. However, this is only a secondary, indirect effect of the ethical value of the rules themselves, and so B does not provide the best answer.

Choice C: Again, it is true that the adoption of the rules by the engineering profession will enhance its image and economic benefits. But like B, this is only a secondary and derivative effect of the rules.

Choice D: The right answer. Since the basic function of the rules is to provide a guide for ethical conduct for engineers, the rules also provide a useful summary for the public of what they have a right to expect from responsible engineers.

EXAMPLE 10.6 The Model Rules of Professional Conduct require registered engineers to conform to all but one of the following rules—which rule is not required?

A) Do not charge excessive fees
B) Do not compete unfairly with others
C) Perform services only in the areas of their competence
D) Avoid conflicts of interest

Solution: There may be one or two problems of this kind in the exam, which can be solved by memorizing the rules or by checking them directly to see which are or are not included. However, it is better to acquire a good understanding of the basic ethical concerns behind the rules, in which case the right answer here will be clear immediately.

Rule A is not required because fees are a matter of free negotiation between engineers and clients. Hence a fee which might seem excessive to some may be acceptable to others because of an interest in extra quality, or unusually quick delivery time, and so on. The other three rules are required.

EXAMPLE 10.7 You are a quality control engineer, supervising the completion of a product whose specification includes using only U.S.-made parts. However, at a very late stage you notice that one of your sub-contractors has supplied you with a part having foreign-made bolts in it—but these aren't very noticeable, and would function identically to U.S.-made bolts. Your customer urgently needs delivery of the finished product—what should you do?

A) Say nothing and deliver the product with the foreign bolts included, hoping this fact won't be noticed by the customer
B) Find (or, if necessary, invent) some roughly equivalent violation of the contract or specifications for which the customer (rather than your company) is responsible— then tell them you'll ignore their violation if they ignore your company's violation
C) Tell the customer about the problem, and let them decide what they wish you to do next
D) Put all your efforts into finding legal loopholes in the original specifications, or in the way they were negotiated, to avoid your company's appearing to have violated the specifications

Solution: Choice A: This is wrong because it is dishonest—it violates the requirement of being objective and truthful in reports, etc.

Choice B: This is wrong because "two wrongs don't make a right." Negotiations with clients should always be done in an ethically acceptable manner.

Choice D: This would violate at least the spirit of the initial agreement. The ethical requirement of being objective and truthful includes an obligation not to distort the intent of any agreements with clients.

Choice C: The correct answer. Being honest with a client or customer about any production difficulties allows them to decide what is in their best interest given the new disclosures, and provides a basis for further good-faith negotiations between the parties.

EXAMPLE 10.8 You are the engineer of record on a building project which is behind schedule and urgently needed by the clients. Your boss wants you to certify some roofing construction as properly completed even though you know some questionable installation techniques were used. Should you:

A) Certify it, and negotiate a raise from your boss as your price for doing so
B) Refuse to certify it
C) Tell the clients about the problem, saying that you'll certify it if they want you to
D) Certify it, but keep a close watch on the project in future in case any problems develop with it

Solution: There are some temptations and half-right elements in some of these, but they must be resisted as not being completely ethical.

Choice A: Even if one informs one's boss of the problems, and negotiates his/her consent to your certifying it, nevertheless it is always wrong to certify work which does not measure up to the highest professional standards.

Choice C: Wrong for the same reason as A. Even if one honestly reveals the problems to the clients, and gets their consent, nevertheless it is the professional duty of an engineer not to certify dubious work.

Choice D: The initial certification would be wrong, no matter how carefully one monitors future progress in the hope of minimizing any future problems.

Choice B: The right choice. Whether or not one's boss is happy with this, it is one's professional duty to refuse certification in such cases, even if as a result one is reassigned or fired.

EXAMPLE 10.9 You are an engineer and a manager at an aerospace company with an important government contract supplying parts for a space shuttle. As an engineer, you know that a projected launch would face unknown risks, because the equipment for which you are responsible would be operating outside its tested range of behaviors. However, since you are also a manager you know how important it is to your company that the launch be carried out promptly. Should you:

A) Allow your judgment as a manager to override your judgment as an engineer, and so permit the launching
B) Toss a coin to decide, since one's engineering and managerial roles are equally important, so neither should take precedence over the other

C) Abstain from voting in any group decision in the matter, since as both a manager and an engineer one has a conflict of interest in this case

D) Allow your judgment as an engineer to override your judgment as a manager, and so do not permit the launching

Solution: Note that a real-life case similar to this problem occurred with the Challenger space-shuttle disaster.

Choice A: Wrong, because engineers have special professional duties and ethical commitments which go beyond those of corporate managers.

Choice B: Wrong for the same reason as A; ethically, engineering responsibilities are more important than managerial responsibilities.

Choice C: Wrong because one's duties as a professional engineer require appropriate action even if other factors may seem to point in the opposite direction or toward abstention.

Choice D: The correct choice. Whatever other duties an engineer has, his/her professional engineering responsibilities must always be given first priority.

EXAMPLE 10.10   Your company buys large quantities of parts from various suppliers in a very competitive market sector. As a professional engineer you often get to make critical decisions on which supplier should be used for which parts. A new supplier is very eager to get your company's business. Not only that, but you find they are very eager to provide you personally with many benefits—free meals at high-class restaurants and free vacation weekends for (supposedly) business meetings and demonstrations, and other more confidential things such as expensive gifts that arrive through the mail, club memberships, and so on. What should you do?

A) Do not accept any of the gifts that go beyond legitimate business entertaining, even if your company would allow you to accept such gifts

B) Report all the gifts, etc., to your company, and let them decide whether or not you should accept them

C) Accept the gifts without telling your company, because you know that your professional judgment about the supplier will not be biased by the gifts

D) Tell other potential suppliers about the gifts, and ask them to provide you personally with similar benefits so you won't be biased in favor of any particular supplier

Solution: Choice B: Wrong, because even if one's company finds such gifts awkward, it is still one's duty as a professional engineer not to become involved in such conflicts of interest.

Choice C: Also wrong. It doesn't matter whether you believe you can remain unbiased, because you still have the conflict, and the possible perception by others that you might be biased by it also remains an ethical problem.

Choice D: Wrong. Remember, two wrongs don't make a right, and the same principle applies no matter how many wrong actions are involved.

Choice A: The right answer. As with the reasoning on choice B, it makes no difference whether others (having less demanding ethical standards than engineers) would find such things acceptable or not. You must not accept any gifts the acceptance of which would involve you in conflicts of interest.

EXERCISES AND ACTIVITIES

Introductory Note

As with any form of ethics, engineering ethics demands more than an impersonal, spectator point of view from us. It is important to consider ethical problems not simply as some other person's problem—to be handled with impartial advice or citing of ethical rules—but instead as a vital problem for oneself which must be resolved in terms that you personally can live with.

To encourage this perspective, many of the cases are presented in the second person, i.e., as problems for you rather than some abstract person. In answering them it is important to maintain this perspective, and honestly explain how you would actually deal with each situation yourself.

10.1 Your company urgently needs a new manager in another division, and you have been invited to apply for the position. However, one of the job requirements is competence in the use of some sophisticated CAD (computer aided design) software. Currently you do not have that competence, but you are enrolled in a CAD course, so eventually you will have that competence (but not before the deadline for filling the new job).

The application form for the job requires you to state whether or not you have the necessary CAD software competence. Is it ethically permissible to claim that you have (in order to get the job for which otherwise you are well qualified), even though strictly speaking this is not true?

Would your answer be any different if your current boss informally (off the record) advised you to misrepresent your qualifications in this way, giving as his/her reason that you are overall the best qualified person for the new managerial position?

10.2 You are a supervisor for a complex design project, and are aware of the NCEES code rule according to which engineers may coordinate an entire project provided that each design component is signed or sealed by the engineer responsible for that design component.

You are very overworked, and so have got into the habit of, in effect, rubber-stamping the design decisions of the subordinate engineers working on your project. You do no testing or checking of your own on their projects, and routinely approve their plans (after they have signed or sealed them) with only the most superficial read-through of their work. In your mind this policy is justifiable because the NCEES rule says nothing about the supervisory responsibilities of someone in your position. Besides which, you trust the engineers to do a good job, and so see no need to check up on them.
Is the supervisor's position here ethically justifiable?

10.3 Newly hired as a production engineer, you find a potential problem on the shop floor: workers are routinely ignoring some of the government-mandated safety regulations governing the presses and stamping machines.

The workers override safety features such as guards designed to make it impossible to insert a hand or arm into a machine. Or they rig up "convenience" controls so they can operate a machine while close to it, instead of using approved safe switches, etc., which requires more movement or operational steps. Their reason (or excuse) is that if the safety features were strictly followed then production would be very difficult, tiring, and inefficient. They feel that their shortcuts still provide adequately safe operation with improved efficiency and worker satisfaction.

Should you immediately insist on full compliance with all the safety regulations, or do the workers have enough of a case so that you would be tempted to ignore the safety violations? And if you were tempted to ignore the violations, how would you justify doing so to your boss?

Also, how much weight should you give to the workers' clear preference for not following the regulations: ethically, can safety standards be relaxed if those to whom they apply want them to be relaxed?

10.4   You and an engineer colleague work closely on designing and implementing procedures for the proper disposal of various waste materials in an industrial plant. He is responsible for liquid wastes, which are discharged into local rivers.

During ongoing discussions with your colleague, you notice that he is habitually allowing levels of some toxic liquid waste chemicals, which are slightly higher than levels permitted by the law for those chemicals. You tell him that you have noticed this, but he replies that, since the levels are only slightly above the legal limits, any ethical or safety issues are trivial in this case, and not worth the trouble and expense to correct them.

Do you agree with your colleague? If not, should you attempt to get him to correct the excess levels, or is this none of your business since it is he rather than you who is responsible for liquid wastes?

If he refuses to correct the problem, should you report this to your boss or higher management? And if no one in your company will do anything about the problem, should you be prepared to go over their heads and report the problem directly to government inspectors or regulators? Or should one do that only in a case where a much more serious risk to public health and safety is involved?

10.5   Your automotive company is expanding into off-road and all-terrain vehicles (ATVs). As a design engineer you are considering two alternative design concepts for a new vehicle, one having three wheels and another the four.

Engineering research has shown that three-wheeled vehicles are considerably less stable and safe overall than four-wheeled ones. Nevertheless, both designs would satisfy the existing safety standards for the sale of each class of vehicles, so either design could be chosen without any legal problems for the company.

However, there is another factor to consider. Market research has shown that a three-wheeled version would be easier to design and produce, and would sell many more vehicles at higher profit margins than would a four-wheeled version.

Is it ethically acceptable for you as an engineer to recommend that your company should produce the three-wheeled version, in spite of its greater potential safety risks? Or do engineers have an ethical obligation always to recommend the safest possible design for any new product?

10.6   Your company has for some time supplied prefabricated wall sections, which you designed, to construction companies. Suddenly one day a new idea occurs to you about how these might be fabricated more cheaply using composites of recycled waste materials.

Pilot runs for the new fabrication technique are very successful, so it is decided to entirely switch over to the new technique on all future production runs for the prefabricated sections. But there are managerial debates about how, or even whether, to inform the customers about the fabrication changes.

The supply contracts were written with specifications in functional terms, so that load-bearing capacities and longevity, etc., of the wall sections were specified, but no specific materials or fabrication techniques were identified in the contracts. Thus it would be possible to make the changeover without any violation of the ongoing contracts with customers.

On the other hand, since there is a significant cost saving in the new fabrication method, does your company have an ethical obligation to inform one's customers of this, and perhaps even to renegotiate supply at a reduced cost, so that one's customers also share in the benefits of the new technique?

More specifically, do you have any special duty, as a professional engineer and designer of the new technique, to be an advocate in one's company for the position that customers should be fully informed of the new technique and the associated cost savings?

10.7   Your company manufactures security systems. Up to now these have raised few ethical problems, since your products were confined to traditional forms of security, using armed guards, locks, reinforced alloys which are hard to cut or drill, and similar methods.

However, as a design engineer you realize that with modern technology much more comprehensive security packages could be provided to your customers. These could also include extensive video and audio surveillance equipment, along with biometric monitoring devices of employees or other personnel seeking entry to secure areas which would make use of highly personal data such as a person's fingerprints, or retinal or voice patterns.

But there is a potential problem to be considered. A literature search reveals that there are many ethical concerns about the collection and use of such personal data. For example, these high-tech forms of surveillance could easily become a form of spying, carried out without the knowledge of employees and violating their privacy. Or the data collected for security reasons could easily be sold or otherwise used outside legitimate workplace contexts by unscrupulous customers of your surveillance systems.

Your boss wants you to include as much of this advanced technology as possible in future systems, because customers like these new features and are willing to pay well for them.

However, you are concerned about the ethical issues involved in making these new technologies available. As an engineer, do you have any ethical responsibility to not include any such ethically questionable technologies in products which you design and sell, or to include them only in forms which are difficult to misuse? Or is the misuse of such technologies an ethical problem only for the customers who are buying your equipment, rather than it being your ethical responsibility as an engineer?

10.8   Can you think of any issue which should be of ethical concern to engineers but is not adequately addressed by current codes of engineering ethics?

If so, should it be added to the current rules, or is it something which cannot adequately be summed up in an ethical rule?

10.9   Can you think of any rule included in current codes of engineering ethics which is unnecessary, too restrictive, or even ethically wrong? If so, should it simply be removed from the current rules, or is it something which serves a useful function even though it is not ethically required?

If you don't think there are any superfluous rules, instead pick one which you think is one of the least ethically important, and explain why you think it is not important.

10.10   As an engineering expert on a state planning board, you have to decide which traffic safety projects (involving installation of traffic lights, road-widening, etc.) should be funded by the state. Previously these matters were decided politically: each region received roughly equal funding, which tended to maximize voter satisfaction and ensure re-election for the politicians. But now non-elected engineering experts such as yourself will decide these matters.

On what grounds should decisions such as these be made? What would be the soundest and most ethically justifiable way for you to allocate the funds?

In particular, would a utilitarian approach (the greatest good for the greatest number) be the best way to allocate funds? But there is a potential problem with this approach. Since the greatest number of people live in cities and suburbs and very few in rural areas, it is likely that scarce funds would always tend to be allocated to cities with this approach, with almost none going to sparsely-populated areas. This policy may tend to save the greatest number of lives or minimize accidents overall, but is it not very unfair to those living in or traveling through rural areas?

But on the other hand, how could you justify spending scarce resources in rural areas when you know that this would benefit far fewer people than if you spent the money in cities instead?

CHAPTER 11

Engineering Work Experience

11.1 A Job and Experience

"How do you get experience without a job, and how do you get a job without experience?"

This is the job-seeker's first, natural question. It was used in a national publicity campaign by the National Commission for Cooperative Education a number of years ago. It's one that all engineering students have considered or will consider at some point during their college career. Career-related experience, commonly called "experiential learning," has become an important supplement to a student's engineering education. Throughout the years, and despite the changing times of up-and-down job markets, numerous employers of engineering graduates, and engineering graduate schools have continued to stress the fact that they seek to fill their annual openings by selecting only from those candidates who have done outstanding things during their college years to distinguish themselves from their peers. In most cases, those students who have participated in some form of experiential learning have gained the types of skills and competencies that prospective employers and graduate schools are seeking.

A college education has evolved over the past century to be quite a different experience from that of our predecessors. For most engineering students during the first half of the 20th century, college was completed in four years. Since studying engineering has always been a rigorous experience, most students chose to devote their time and efforts to experiences in the classroom, library, or laboratory. Those who chose to work while going to school were viewed by the conventional student population as "nontraditional," representing the "working class" who were forced to work for pay solely as a means of affording their schooling. However, in the latter part of the 20th century the costs of a college education soared. It is now the rule, rather than the exception, that most students are forced to work for pay in order to keep up with steadily increasing tuition, room and board, and associated expenses such as books, computers, and transportation. (Again, some countries fully subsidize college education—with one of the benefits being that students can concentrate fully on their studies, with less need for extracurricular employment during school.)

For many students, the idea of combining their academic studies with a career-related employment experience is something they had not previously considered. The thought of stretching their college education to five or even six years can seem to be an unnecessary burden. To their parents who grew up with the notion that a college education is a four-year plan with a direct path into the world of employment, lengthening the time spent in school is also an uncomfortable concept.

Consider the change in employment trends in recent years. The current generation of college students was born, for the most part, in the mid-1990s. During the relatively short span of their lifetime, this group has witnessed firsthand several major shifts within corporate America. Prior to the years when many of these students were born, a major recession was experienced in some of the largest manufacturing sectors. Especially hard hit was the automobile industry and the surrounding Midwest region of the country. Dubbed the "Rust-Belt," this area saw a significant reduction in the work force, especially in the traditional "blue-collar" jobs. Jobs that had traditionally been viewed as secure for a worker's lifetime were now eroding and in many cases disappearing. During this period, the employment of professionals remained fairly stable, and new engineering graduates were having only minor problems gaining employment at graduation time.

The period 1983–1986 saw a revival in the U.S. economy and this was reflected by an increase in hiring of new engineers, with salaries well above the rate of inflation.

The period 1988–1994 was one of major restructuring within corporate America. Many major firms instituted a series of layoffs and experienced periods of downsizing. This period was notable for how the "white-collar" work force was affected. Professionals either were being laid off or provided with incentives to retire early. The net effect was a significant restructuring, reflected by one of the weakest labor markets ever for engineers.

The Collegiate Employment Research Institute at Michigan State University conducts an annual hiring survey of employers and is able to track the labor market for new college graduates. From 1994 to mid-2001, the job market for graduating engineers was excellent; experts agree that it was probably the hottest job market in over 30 years. Unfortunately, overall hiring contracted nearly 50% in 2002–2003, due in part to the aftershock of 9/11. Hiring trends slowly recovered each year from 2004–2008, but then the economic downturn of 2008 began to negatively influence employment opportunities for new college graduates, including engineers, as available positions became highly competitive. It was not until 2010–2011 when a gradual shift in hiring patterns began to emerge. Students were once again finding that employers were returning to college campuses, job prospects were slowly improving, and more graduates were finding employment.

The most recent survey of over 4,300 employers for 2012–2013 shows a slowly growing economy and a continued tight labor market for new college graduates. If this trend continues, opportunities for engineers will still be available, but they will require that students devote increased efforts to finding the right opportunity. However, it should be noted that employment opportunities and starting salaries for engineers have traditionally surpassed those of most other majors on college campuses, and this trend continues.

Truly, the current generation of college students has witnessed a roller-coaster type of economy. Many realize that the rules of the game have changed and that lifelong job security with a single employer is a thing of the past.

However, one fact remains clear from employers in good times and bad: the need for college graduates with experience is a constant. Many employers expect today's college graduates to have significant work experience. Why? Because engineering graduates with career-related work experience require less training and can produce results more quickly than the typical recent college graduate. To an employer, this is sound economic policy.

There are many forms of experiential learning available to engineering students. This chapter will discuss several of the most popular options, including on- and off-campus jobs, summer work, volunteer experiences, academic internships, research assistantships, and perhaps the most popular and beneficial program for students, cooperative education. Each of these has its own advantages and disadvantages, depending on a student's particular situation and background. These issues will be examined as part of this chapter.

11.2 Summer Jobs and On- and Off-Campus Work Experiences

Most engineering students arrive at college with some type of work experience. Some have mowed lawns, done baby-sitting, or worked in fast-food franchises, while others may have had the opportunity to actually work in a business or industry that was involved in some type of engineering activity. Whatever the experience, it can be helpful in preparing you for your future. Career-related work experience is clearly preferred by engineering employers, but one has to start someplace in developing a work history.

There is value in working in the residence hall cafeteria, or delivering pizzas, or helping out in one of the campus offices. These experiences begin to build a foundation which your classroom learning and other career-related experiences will supplement. On- and off-campus jobs and summer work experiences may not seem highly relevant, but they can help develop important skills. Through such experiences, many students have strengthened their communication skills, learned to work as part of a team, and developed problem-solving abilities and an organizational style that can be applied to engineering courses and future employment positions. Jobs that you hold as a student will influence your perceptions about the things you like to do, the conditions and environments in which you like to do them, and the types of people and things with which you like to work. These jobs can be useful in helping you identify your personal strengths and weaknesses, interests, personal priorities, and the degree of challenge you are willing to accept in a future position.

"Real world" experience can come in many forms. While career-related experience is important, do not underestimate the importance of learning some basic "on the job" skills. If you are seeking to develop some basic job skills or merely trying to earn some extra money to assist with your college expenses, you should not be afraid to explore some of the many employment options that are currently available on and off campus during the school year and in the summer.

11.3 Volunteer or Community Service Experiences

Another popular form of experiential learning is found in the many opportunities associated with volunteer or community service experiences. While these programs are available to engineering students of all class levels, they can be particularly beneficial to freshman or sophomore students who are trying to gain some practical experience. Generally, volunteer or community service projects can be short- or long-term arrangements. Typically, these positions are with nonprofit organizations such as human service groups, educational institutions at all levels, a variety of local, state, and federal government agencies, or small businesses. These groups are particularly interested in engineering students for their strong backgrounds in math, science, and computer skills, which can be applied to a variety of projects and activities.

Even though these positions usually are not paid, students can gain significant experience and develop important skills. Students have found that working with different groups can enhance their communication abilities, as they are often interacting with people much different from those in their collegiate environment. Through the various projects and activities, students can develop organizational skills, initiative, and the desire for independent learning that can be useful in many settings including the classroom and other employment situations. Students involved in volunteer and community service experiences find these opportunities helpful with career decisions, while building confidence with the satisfaction of helping others and providing needed services.

For those students interested in developing a basic set of skills which can be useful throughout their professional engineering career, volunteer and community service work is worth exploring. For more information, contact your campus placement center or a particular firm or agency in your area. A faculty member or academic advisor may be able to provide assistance in this very valuable endeavor.

11.4   Supervised Independent Study or Research Assistantship

The *Supervised Independent Study* or *Research Assistantship* is a form of experiential learning which is designed primarily for the advanced undergraduate engineering student who is considering graduate school and/or a career in research and development. Generally, a supervised independent study or research assistantship is a planned program of study or research that involves careful advanced planning between the instructor and student, with the goals, the scope of the project, and evaluation method specified in writing. Some of these opportunities are paid, while others involve the award of credit, and possibly both.

For students considering graduate school or a career in research and development, these opportunities can provide experience working in an environment which can be quite different from other engineering functions. Graduate school and research work involve an application of technical skills at a very theoretical level. Students should enjoy the challenges of advanced level problem solving which will require a very in-depth use of their engineering, math, and science background.

To learn more about supervised independent study programs or research assistantships, talk with some of your professors whom you enjoy working with and with whom you share similar engineering interests.

11.5 **Internships**

What is an internship? Of all forms of experiential learning, some think that this is the most difficult to describe, as it can have different meanings, uses, and applications depending on a variety of circumstances. Generally, *internships* can be either paid or unpaid work experience that is arranged for a set period of time. This usually occurs during the summer so there is the least disruption to a student's academic schedule. For most students and employers, an internship is a one-time arrangement with no obligations by either party for future employment. Objectives and practices will vary greatly depending on the employer and the student's class level in school. Some internships involve observing practicing engineers and professionals to see what working in the field is "really like." The intern's actual work often involves menial tasks designed to support the needs of the office. However, other internships can be structured as a "capstone experience" which permits students to apply the principles and theory taught on campus to some highly technical and challenging real-world engineering problems.

At most engineering schools, internships are treated as an informal arrangement between the student and the employer. Therefore, this activity is usually under the general supervision of an experienced professional in the field in a job situation which places a high degree of responsibility on the student for success or achievement of desired outcomes. While many campus placement centers provide facilities and assist with arrangements for internship interviews, there generally is not a formal evaluation of the internship job description by any engineering faculty member. As a result, it is often difficult for students to understand how, and whether, this internship will be related to their engineering field of study. Since the typical engineering internship does not involve faculty input, there is usually not any formal evaluation or monitoring of the experience by any college officials. For these reasons, college credit is usually not awarded for these experiences.

For many students, however, an internship can be a very valuable part of their college education. Under the best of circumstances, the internship will be structured by the employer so it will be directly related to the student's field of study. It is advantageous for the student to try to obtain, in advance, a description of the duties and responsibilities of the position. The student may wish to review this material with an academic advisor or faculty member to gain an opinion on the relevance to the student's selected field. These individuals may be able to make suggestions and comments on ways that the student can maximize the benefits of the experience. The most productive internships are obtained by upper-level undergraduates who have completed a significant amount of engineering course work. It is often easier for employers to match these students and their qualifications to specific projects and activities which will take advantage of the student's academic experiences. Students gain some practical real-world experience which will provide them with a better perspective of the engineering field that has been selected. For the employer, internships provide a mechanism to supplement their workforce in order to complete many short-term engineering projects. It also enables the employer to evaluate students in a variety of work-related situations. Many employers will use internship programs to screen candidates for possible full-time employment after graduation.

For students with restrictions related to time constraints, curriculum flexibility, scheduling difficulties, location or geographical preferences, the one-time internship program can serve as a valuable supplement to their engineering education. While internships do not provide the depth of experience found in other forms of experiential learning, they can help students gain an appreciation for the engineering profession and perhaps provide additional opportunities for full-time employment after graduation.

11.6   Cooperative Education

At most engineering schools, cooperative education programs are viewed and promoted as the preferred form of experiential learning.

> *Engineering cooperative education (co-op) programs integrate theory and practice by combining academic study with work experiences related to the student's academic program. Employing organizations are invited to participate as partners in the learning process and should provide experiences that are an extension of and complement to classroom learning.*

This background statement, endorsed by the Cooperative Education Division of the American Society of Engineering Education, describes the basic foundation of *cooperative education* and some of the critical components of a quality engineering co-op program. Perhaps the most important part of any cooperative education program is its linkage to a college's engineering programs. Co-op is considered to be an academic program, and in fact at many colleges of engineering it is administered centrally, by the college itself.

As an academic program, co-op is structured so student assignments are directly related to their major field of study. Typically, participating employers, who have been previously approved by the program administrators will forward detailed job descriptions, including salary information of available positions, to co-op faculty and staff. These descriptions are reviewed to evaluate their correlation to the academic content and objectives of their educational program. Employers also provide a set of qualifications for the candidates they are seeking to provide appropriate matching of students with the positions they have available.

The integration of school and work is another important component of cooperative education programs. This is typically achieved by alternating several periods of increasingly complex co-op assignments with periods of study at school, almost always with the same employer. This program is called the "*alternating-term co-op*" and is the most common form of co-op scheduling. For example, a student might be at the XYZ Company during the fall semester, back at school for the spring semester, then return to XYZ for the summer semester, back to school for the fall, etc. This type of schedule allows students to apply classroom learning to real-world problems in an engineering setting. Upon returning to school, students are then able to relate the learning in the classroom to the experience gained on the job. After another semester of schoolwork, students return to the employers with an even greater knowledge of engineering and scientific principles, which can be used for even more complex assignments in the workplace. As students mature academically and professionally they become stronger students and valuable employees. It is truly a winning combination for the employers, the students, and the schools.

Another form of cooperative education scheduling which is popular at some colleges and universities is called "parallel co-op." Under this arrangement, students usually enroll for a partial load of class work while engaging in co-op assignments for 20 to 25 hours per week. This type of co-op program allows students to continue to make progress toward their degrees while gaining valuable work experience. The disadvantage of this arrangement is that students may encounter difficulties if the demands of the co-op position begin to interfere with the time needed for school assignments, or if the time devoted to schoolwork begins to hamper the performance on the co-op assignments. Great care and judgment should be exercised when students and employers consider undertaking this co-op option.

Under another arrangement, co-op assignments are scheduled consecutively, during back-to-back semesters. This option is particularly useful for co-op projects that require a longer commitment than a single semester, or where the activity is seasonal in nature. For example, many civil engineering students are employed by highway construction firms for assignments that typically last for a summer and fall semester. A student involved in this type of co-op may work two consecutive semesters, followed by a semester or two in which they attend school and then a return to the employer for one or two additional co-op assignments. This pattern can provide many of the same benefits of the alternating plan if it is scheduled in a way which ensures that the academic course work is still a major component of the overall plan.

Some co-op schedules are actually a combination of these different patterns. Due to the sequential scheduling of a typical engineering curriculum, it can be difficult to adhere to a true alternating co-op program. Therefore, some students and employers will develop a series of assignments that may begin as an alternating plan for the first two or three rotations, but which may then require a back-to-back schedule or a parallel semester to complete the full schedule plan due to a student's particular academic schedule.

Cooperative education is rarely a summer-only program. Interrupting work for two semesters every year makes it difficult to maintain continuity in the integration of academics and the employment experience. In addition, most participating employers have needs which require them to have co-op students available for these positions throughout the year. In fact, most co-op employers plan their hiring so that while one student is away at school, another student is on-site working on a given co-op assignment.

In order for both an employer and a student to gain maximum benefit from the co-op experience, most employers and schools require that a student make a commitment for three to four semesters. This guarantees stability for the employers while ensuring the student at least a full year of job-related experience. Most engineering schools typically will recognize the student's participation in co-op after completion of the equivalent of one year of work experience by placing an official designation on the student's transcript and/or diploma.

Advantages of Cooperative Education Programs

. . . Advantages for Students

Cooperative education programs provide students with an advantage for employment or graduate school consideration. Many employers seek to fill permanent positions with students who have had significant work experience. These employers use the cooperative education program to evaluate candidates on the job. For many students this usually leads to an offer of permanent employment upon graduation. Students are under no obligation to accept full-time employment from their co-op employer, but studies have demonstrated that many will accept these offers. Students who choose to work for their co-op employer often begin with higher seniority and benefits, including larger starting salaries, increased vacation time, and other fringe benefits. Those students who elect to interview with other firms usually find that their co-op experience places them far ahead of their peers, as they are sought more often for interviews and plant trips, and receive more offers, and earn higher starting salaries than their classmates.

Students who have participated in co-op programs usually comment on how they have been able to develop their technical skills. The nature of co-op assignments allows students to use the skills gained in the classroom and apply them to actual engineering problems. In the classroom, students' assignments and laboratory experiments are graded based on how close they come to achieving the correct answer. As they quickly learn on the job, the real-life assignments and projects often are not directed toward a specific answer, but a variety of potential solutions. This is where the student's technical background is developed and applied. Much of the actual technical learning occurs when students are forced to apply various principles to these unique situations. The learning that is achieved as part of the co-op experience provides students with new knowledge that often will make them better prepared for classroom challenges. The opportunity to "learn by doing" has been an important component of human development throughout time.

Many students who participate in cooperative education programs also find that the experiences can help solidify career decisions. Many engineering students are not fully aware of the spectrum of options and opportunities available to them. By engaging in the daily responsibilities of engineers, interacting with both the technical and non-technical individuals that work with engineers, and applying their academic background to their positions, students begin to gain a better understanding of the engineering profession. As a result, many find that their selection of a particular discipline has been confirmed, reinforced, and even strengthened. Others find that they have been exposed to new fields and technologies of which they had not previously been aware. Some change majors to reflect their new interests, while others modify their course schedule to include new courses to incorporate some of the exciting new areas to which they have been exposed.

All cooperative education assignments are, by definition, paid positions. Therefore, students find that the salaries they earn as co-op participants can greatly assist them in meeting the high costs of education. Recent studies have demonstrated that average salaries paid to engineering co-op students are in the neighborhood of $2,400 per month. For students who are working between four and seven months per year, co-ops can provide significant resources to meet their education expenses. It should be noted that average salaries can vary greatly depending on such factors as the employer, geographical location (pay is often higher in high-cost-of-living areas), the students' majors, class levels, and the amount of experience they have had.

. . . Advantages for Employers

Many students often wonder why employers become involved in cooperative education programs. Some of the most common reasons given by participating employers are:

- The cost of recruiting co-op students averages sixteen times less than recruiting college graduates.
- Almost 50% of co-op students eventually accept permanent positions with their co-op employers. The retention of college graduates after five years of employment is 30% greater for co-op graduates.
- Typically, co-op students receive lower salaries and fewer fringe benefits than permanent employees. Total wages average 40% less for co-op students. In addition, employers are not required to pay unemployment compensation taxes on wages of co-op students if they are enrolled in a qualified program.
- The percentage of minority members hired is twice as high among co-op students as among other college graduates, thus assisting co-op employers in meeting EEO objectives.
- Co-op programs provide an opportunity to evaluate employees prior to offering them full-time employment.
- The co-op graduate's work performance is often superior to that of a college graduate without co-op experience. Co-op students are typically more flexible, and easily adapt to a professional environment.
- Regular staff members are freed up from more rudimentary aspects of their jobs to focus on more complex or profitable assignments.
- Co-op programs often supply students who have fresh ideas and approaches and who bring state-of-the-art technical knowledge to their work assignments.
- Co-op graduates are capable of being promoted sooner (and farther) than other graduates.
- Co-op programs build positive relationships between businesses and schools, which in turn helps employers with their recruiting.

. . . Advantages for Schools

There also are many benefits to educational institutions that offer cooperative education programs to their engineering students.

- Cooperative work experiences provide for an extension of classroom experience, thus integrating theory and practice.
- Cooperative education keeps faculty members better informed of current trends in business and industry.

- Co-op programs build positive relationships between schools and businesses, and provide faculty members with access to knowledgeable people working in their fields.
- Co-op programs enhance the institution's reputation and attract students interested in the co-op plan to their school.
- Cooperative education provides schools with additional business and industry training facilities that would be unaffordable otherwise.
- Cooperative education lowers placement costs for graduates.

More Benefits of Co-op

Co-op students have found that many professional development skills so critical for success in today's workplace can be enhanced through their co-op experiences.

1) **Written and oral communication skills:** The ability to communicate ideas, both in written form and orally, is a skill that is critical. Most co-op assignments require students to document their findings and report them to others. This may take the form of an e-mail to a supervisor, a technical memo, a presentation to a group of engineers and technicians, or even a formal presentation to a group of directors or other executives. While communications and presentations may not be a favorite activity for many engineering students, co-ops provide students the opportunity to improve and develop these critical skills. For those who find this communication a liability, cooperative education assignments can help turn it into an asset.

2) **Networking:** Most co-op students find the opportunity to develop new contacts and work with people from a variety of backgrounds a great asset to their short- and long-term career objectives. Co-op students come in contact with many individuals who are able to provide advice, lend support with projects, and help get things done more efficiently. These same people can recommend new assignments to co-op students, and maybe even write an important letter of recommendation.

3) **Self-discipline:** Most co-op students quickly discover that the transition from school to work requires that they develop an organizational style which may be superior to that used in college. They find that within the co-op structure they are given a great amount of responsibility and freedom to succeed (or fail) with their projects. Time management, punctuality, adequate preparation, organization, and specific protocol and procedures are learned on the job. As they acquire and develop these skills, they find that they are able to apply them back at school, which in turn makes them more productive and successful students.

4) **Interactions with a variety of people and groups:** Co-op students deal with a wide range of individuals as they complete their tasks. Typical individuals may include engineers, technicians, scientists, production workers, labor union representatives, clerical staff, managers, directors, and perhaps even CEOs. Each of these will contribute to the students' learning as they experience the successes, challenges, frustrations, and friendships that will evolve from these interactions. These individuals can help the co-op student evolve from a student to a successful engineer.

5) **Supervisory and management experience:** Many cooperative education assignments require that the student take responsibility for other individuals or

groups in order to accomplish their tasks. These opportunities to manage the work of others can provide invaluable experience to students, especially those who aspire to eventual management or leadership positions.

A Note of Caution

As discussed above, cooperative education programs can provide many positive benefits to students, employers, and schools. However, a co-op program may not be the appropriate form of experiential education for every engineering student. There are many aspects to the co-op structure that may not be consistent with a particular student's overall goals and objectives.

The biggest drawback to co-op participation is the intermingling of work and study. During each of the semesters that students are off campus working on their co-op assignments, they are not making progress toward their degree requirements. This means that typical co-op students add an extra year to the duration of their academic programs. For many, this means that co-op participation will extend their engineering degree program to five or even six years. The overwhelming majority of co-op students, schools, and employers feel that this sacrifice is minor compared to the outstanding benefits received from this type of education. However, students and parents should carefully weigh this factor.

There are other factors which also should be considered. The continual relocation between school and the work is a minor inconvenience, but one which takes time and effort—especially if the two sites are a substantial distance apart.

The engineering curricula also can present a difficulty due to the nature of the sequential offering of certain classes and their prerequisites. Students should work out a long-term schedule with their academic advisor and co-op coordinator so that any class scheduling problems can be handled.

To find out more about engineering cooperative education programs at your school, contact your co-op office or placement center, or talk with your academic advisor or a faculty member. If your school doesn't have a co-op program, you may want to write or talk with a corporation of interest to you. You may be able to work out a co-op assignment with the help of an advisor, or faculty member, or the campus career center.

11.7   Which Is Best for You?

Many forms of experiential learning have been discussed in this chapter. Each of these opportunities has advantages and disadvantages associated with it. Only you can determine which program is best suited to your particular situation. Begin by asking yourself these types of questions:

- Is obtaining career-related work experience a high priority in my educational planning?
- Am I willing to sacrifice personal convenience to gain the best possible experience?

- Will I be flexible in considering all available work opportunities, or do I have special personal circumstances (class schedules, time constraints, geographic or location restrictions) which limit my choices?
- Do I have the drive and commitment necessary to succeed in a co-op program?

Your answers to such questions will be important as you move forward in your engineering career. Work with your academic advisor, a professor, or your campus placement office to get as much information as possible to make an informed choice in determining which is right for you. The resources are available to you—it is up to you to take advantage of them.

EXERCISES AND ACTIVITIES

11.1   Develop a list of your skills and abilities. Determine which could be improved with work experience and which will be developed in the classroom.

11.2   Interview one of your professors who is involved with a research project. Discuss the advantages and disadvantages of a career in research.

11.3   Talk with other students who have done volunteer and community service projects. List the benefits they have gained which have helped them in their engineering classes.

11.4   Visit your engineering co-op office or campus career planning or placement center and read the recruiting brochures of some of the employers in which you have interest. Make a list of the types of qualifications these firms are seeking for full-time and co-op or intern positions.

11.5   Attend a co-op orientation session. List some of the advantages and disadvantages as they relate to your personal priorities.

11.6   Attend your campus career/job fair. Talk with at least three recruiters from different firms. Write a report comparing their views of cooperative education experiences, internships, and other forms of experiential learning.

11.7   Visit with some upper-class engineering students who have been co-ops or interns. Write an essay that compares and contrasts their experiences. Which students do you feel gained the most from their experience, and why?

11.8   Visit the campus placement center. Gather information concerning the starting salaries and benefits of those students who have participated in cooperative education compared to those who have done internships or some other form of experiential learning.

11.9 Visit the engineering co-op office and read some of the job descriptions of current openings. Select one of these descriptions and write an essay discussing the benefits you feel would be gained from this position.

11.10 Meet with your academic advisor and develop a long-range course schedule. Modify this plan so it shows how a co-op and an internship would affect this schedule. Write a short paper discussing the advantages and disadvantages of extending your education to include extra time for work experience.

11.11 Develop a list of your personal career priorities. How can this list be enhanced by some form of experiential learning?

11.12 Develop a list of your personal and educational abilities that you think need to be strengthened. Write an essay that discusses how work experience could improve these skills for you.

11.13 Interview an engineering professor or graduate student who had been a co-op student as an undergraduate. Discuss how the co-op experience influenced them to pursue an advanced degree in engineering.

11.14 Visit <http://epics.ecn.purdue.edu> and review the volunteer student team projects that have been developed at this university. Prepare a short paper that discusses the benefits gained by these students and how these experiences can be an advantage in academic studies.

11.15 Attend a campus career fair. Talk with at least three recruiters about their opinions concerning career-related work experience. Develop a report that discusses the reasons that employers prefer to hire students for permanent positions who have had significant work experience as part of their undergraduate education.

11.16 Talk with an advanced engineering student who changed their engineering field of study as a result of a co-op or intern work experience. What factors in their experience caused them to change their major? Do they view this work experience as a disadvantage or as an advantage to their overall career planning?

Nine Excel® Skills Every Engineering Student Should Know

Pramod Abichandani

CONTENTS

A.1   Turning Data into Information: Deepwater Horizon and Wall Street

In 2010, the Gulf of Mexico witnessed one of the largest accidental marine oil spills in the history of the petroleum industry (Fig. A.1). The Deepwater Horizon oil spill lasted for approximately three months. An explosion at the Deepwater Horizon semi-submersible Mobile Offshore Drilling Unit (MODU), which was owned and operated by the company Transocean, caused the spill. The Deepwater Horizon was operating in the Macondo Prospect. The British multinational oil and gas company British Petroleum (BP) is the developer of the oil fields in the region.

In 2011, BP was one of the largest energy companies in the world in terms of revenue. Its stocks are listed on the London Stock Exchange and the New York Stock Exchange (NYSE). A quick look at BP's stock price in 2010 reveals an interesting picture. Figure A.2 illustrates that BP's stock closing price in the months between January and May fluctuated between $52.4 and $62.32. However, the explosion on April 20, 2010, initiated a dramatic drop in the closing price of BP stock. The stock price steadily declined and hit a minimum of $27.02 on June 25, 2010. This is the exact date when Hurricane Alex was formed, causing a delay in the efforts to check the spread of oil. The subsequent

Figure A.1 An image of the lingering oil slick caused by the oil spill off the Mississippi Delta on May 24, 2010, captured by the Moderate Resolution Image Spectrodiometer (MODIS) on NASA's Terra satellite.

Source: NASA.

Figure A.2 Closing price of BP stocks listed on the NYSE during 2010.

weeks saw BP making major strides in its efforts to stop the oil spill. This was reflected in the steady but often fluctuating rise of its closing prices during the second half of 2010. This graph reveals the complicated nature of contemporary technology, business, and engineering relationships.

The changing technology, business, and engineering climate has resulted in an increasing number of Science, Technology, Engineering, and Mathematics (STEM) professionals, especially engineers, across all disciplines working with large datasets that are generated as a result of real-world operations. Financial market data, environmental pollution and temperature data, data collected by medical devices, cell phone network quality data, highway congestion data, air traffic data, data from social networks, sensor data for cellphone apps, and voting data are a few examples of such large-scale datasets. Such large datasets exhibit diversity in terms of their sources, are longitudinal in nature, and are referred to as BigData by the STEM community, including the National Science Foundation (NSF). In this data-driven economic climate, the success of many commercial, research, and governmental organizations depends on the data-analysis capabilities of their engineers.

Organizations, aware of this fact, are increasingly looking for STEM professionals who can critically evaluate and present data using modern software applications and frameworks like MATLAB, Excel, R, SAS, SPSS, and Apache™ Hadoop™. More importantly, these STEM professionals are required to synthesize quantitative analyses into qualitative information that can drive decision making for policy development, financial planning, healthcare management, and organizational strategies. These decisions, some of which are highly controversial, have profound effects on businesses, the environment, lifestyles, and our standard of living.

A.2 Data Analytics: Goals and Core-Connector Concepts

This goal of this chapter is to immerse students in the technical challenges of engineering data analytics. In general, data analytics is defined as the process of procuring data generated by real-world operations from various sources such as sensors, worksheets, software database, and websites; analyzing these data using descriptive and predictive analysis techniques; and using these analyses to generate independent insights and present them in a visually effective manner. The ultimate aim of data analytics is to facilitate informed and data-driven decision making. These core-connector concepts remain common across most tasks and are illustrated in Figure A.3.

Throughout this chapter, the technical details of these core-connector concepts will be explicated. Examples that span various engineering disciplines will be discussed, including applications in the areas of stock markets and environmental engineering. Of main importance is the knowledge of mathematical concepts required to analyze the data at hand. Equally important is the use of a software application to implement the required mathematical analyses and generate compelling visualizations.

2.1 The Nature of Data

The *Oxford English Dictionary* defines "data" as facts and statistics collected together for reference or analysis. True to this definition, data will be used in this chapter to perform meaningful analyses and generate insights. The data used in this chapter will be open source (i.e., data that are available freely for public use). These data will be sourced from

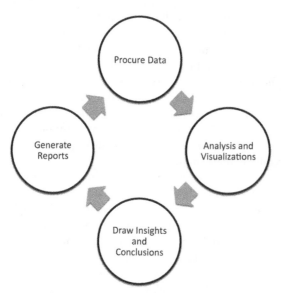

Figure A.3 The four core-connector concepts of data analytics. Successful analysts are able to perform all the tasks involved in these four core-connector concepts.

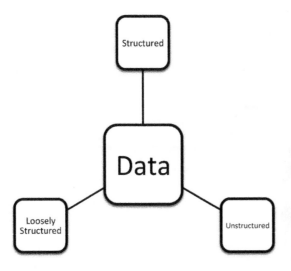

Figure A.4 Data can be classified in three general categories depending on the nature of the engineering process that generates the data and the method in which data are captured.

websites, including Google/Yahoo Finance (for stock market information), the U.S. Environmental Protection Agency (EPA), and other organizations.

In the most general sense, data fall into one of the following categories (Fig. A.4):

1. Structured Data: As the name suggests, "structured data" refers to data that are organized in a structure and hence can be identified. The schedule of your classes is an example of structured data since it can be completely specified in terms of day, date, month, year, location, and course number. The NYSE stock information of BP stored in the file BP_NYSE2010.xls is an example of structured data since it is stored in terms of the open price, high price, low price, close price, and volume of stocks traded on a given date (Fig. A.5).

Open	High	Low	Close	Volume
57.97	57.97	57.97	57.97	0
59.3	59.45	59.08	59.15	3956568
59.65	59.9	59.31	59.57	4109563
59.52	59.92	59.34	59.88	6230175
59.92	60	59.69	59.86	4432171
59.79	60.06	59.67	60	3786118
61.54	61.97	61.45	61.89	8229519
60.86	61.69	60.8	61.5	6127216
61.31	61.98	61	61.8	5688621
61.22	61.75	61.16	61.73	3330668
61.64	61.64	61.64	61.64	0
61.64	61.64	61.64	61.64	0
61.76	62.38	61.76	62.32	4914423

Figure A.5 The NYSE stock information of BP stored in the BP NYSE2010 .xls file is an example of structured data.

Figure A.6 Examples of unstructured data include Mark Zuckerberg's Facebook post about a new Facebook campus and Google's Tweet about procuring more renewable energy for data centers.

2. Loosely Structured or Unstructured Data: "Loosely structured data" or "unstructured data" refers to data that are not organized in any particular manner—for example, the data available on social networks such as Facebook and Twitter. A Facebook post can consist of a variety of data types such as text, images, website links, videos, etc. (Fig. A.6). Similarly, a 140-character-long tweet can consist of text in any language and may include links.

2.2   Why Microsoft Excel?

Typically, most data analytics tasks start with procuring relevant data. An engineering data analyst must ascertain where to get the datasets best suited for the task at hand. Additionally, he/she must use the software application best suited for the task at hand. There are occasions where engineering data analysts must record datasets by manually entering data in the analytics software system. There are several software applications and programming languages that can be used to perform analytics; examples are illustrated in Figure A.7.

Microsoft Excel will be used as the primary software tool to handle data in this chapter. Excel was developed and released by the software company Microsoft in the 1980s and to date remains one of most widely used tools for recording, analyzing, and sharing data. Excel is made available by Microsoft Corp. as a part of its Office

Figure A.7 Some of the popular software applications and programming languages used for data analytics—Python, R, Matlab, Excel, SAS, SPSS, Ruby, Java, Xcode, Microsoft Visual Studio.

suite of software applications. Office suite is widely available in most computers across universities.

While other applications such as SPSS and SAS feature advanced functionalities for data analytics, Excel provides an easy way to get started with data analytics tasks for beginners. SPSS in particular has a spreadsheet interface very similar to Excel (Fig. A.8, top). Once students are familiar with the Excel environment, switching to the SPSS environment is a simple process. SAS has an interface that allows users to programmatically implement data analytics tasks (Fig. A.8, bottom).

A typical Excel file looks like a sheet of rectangular cells arranged in a tabular format. This type of arrangement of rectangular cells is called a spreadsheet (Fig. A.9). All data in Excel are organized as a spreadsheet of cells, with each cell being referenced by a combination of column and row number respectively.

A cell is the basic organizational unit of a spreadsheet. A cell can be thought of as a rectangular box that holds a value or a formula. The value can be a number or text. Alternatively, a cell can be left blank. By default, an Excel file, also known as an Excel Workbook, is stored using a .xlsx extension. A new Excel Workbook has one spreadsheet, but more sheets can be added by using the + symbol at the bottom of the Workbook (Fig. A.10).

A.3 The Nine Excel Skills

We are now ready to dive right into the technical challenges of data analytics. We will use Microsoft Excel to perform all the required mathematical analyses. As illustrated earlier in Figure A.3, performing data analytics involves the following steps:

A

B

Figure A.8 SPSS, an IBM software application, is widely used for data analytics. It features a spreadsheet interface similar to Excel (source: http://upload.wikimedia.org/wikipedia/en/3/33/Spss-19-screenshot.jpg). SAS is also widely used for data analytics. SAS provides a host of programming capabilities to perform data analytics tasks (source: http://www.sas.com/technologies/bi/appdev/base/index.html). Excel provides an easy way to get started with data analytics for beginners. Once you are proficient at performing data analytics tasks in Excel, switching to more advanced data analytics software applications such as SPSS and SAS is relatively simple.

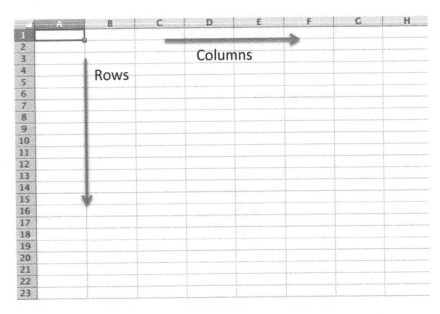

Figure A.9 An Excel Spreadsheet consists of rectangular cells arranged in a tabular format. Each cell can be referenced by a combination of column and row number respectively. The first cell is referenced as A1.

1. Procuring data generated by real-world operations from various sources such as sensors, worksheets, software database, and websites
2. Analyzing these data using descriptive and predictive analysis techniques, and visualizing this analysis
3. Using these analyses to generate independent insights
4. Generating reports that synthesize the analyses in a visually effective manner

Figure A.10 Additional sheets can be pressing the + symbol on the bottom of the sheet.

We have identified nine Excel skills that allow engineering students to perform effective data analytics. In the following discussions, we will describe in details each of these skills using data from several sources.

3.1 **Entering Data Manually**

You can manually enter data in an Excel spreadsheet by simply typing the data in the cells. To perform this task, you start by double-clicking on the Microsoft Excel icon and starting a new blank Excel file. A recommended Excel practice is to provide a name to the column in which the data is being entered. As shown in Figure A.11, we enter numbers from 0 to 1 in increments of 0.1 and name the column X1. Similarly, two more columns, X2 and Y, are entered manually in this Excel spreadsheet. Note that the column Y has entries that are squares of entries in column X2.

3.2 **Importing Data**

You can also import large datasets automatically in Excel. To illustrate this, we will start by revisiting the BP stock performance in 2010. To get the data for BP stock performance in 2010, we make use of publicly available online sources such as Google Finance or Yahoo Finance. These websites provide up-to-date and relevant market data.

You can procure information about a particular stock by entering the corresponding ticker information in the search bar located on the top of the Google Finance page. A ticker is a unique identifier that is assigned to each stock that is traded via an exchange. It usually consists of an arrangement of letters. For example, the ticker for British Petroleum is BP and that of Apple is AAPL.

On entering the ticker for British Petroleum (BP) in the search bar on the top of the Google Finance page, the market information about BP's stock shows up. As shown in

Figure A.11 Manually creating an Excel spreadsheet by entering data.

Figure A.12, the standard Google Finance page allows you to view the current and historical market performance of a particular stock. On the left side of the webpage, there is a link that allows you to select historical daily prices.

For this analysis, we are interested in the prices from January 1, 2010, to December 31, 2010. Navigate to the Historical prices page. Once this range of dates is entered and the stock price is updated to reflect this range, you can download the entire dataset to a spreadsheet using the Export functionality provided (Fig. A.13). The downloaded spreadsheet should be saved to a known location on the computer.

This spreadsheet can be imported into Microsoft Excel using the Import option in the File Menu. On selecting the import option, a window similar to the one shown in Figure A.14 appears. Select a file format. In this case, select the CSV file option since Google Finance provides a .csv file when data are downloaded to a spreadsheet.

Following this step, the Text Import Wizard appears (Fig. A.15). The Text Import Wizard allows you to import a file containing data using the following three steps:

1. Choose the file type that best describes the data. The file can be a delimited file or a fixed-width file.
2. In the next step, the Text Import Wizard provides options to select the delimiting character.

Figure A.12 Use of Google Finance website to get information about BP NYSE stock prices.

3. The final step allows you to provide information about the format (e.g.,: date, text, etc.) of each column in the imported file.

Since commas delimit the data in the BP stock prices file, the delimited option is chosen in step 1, as shown in Figure A.15.

A

Show: Daily | Weekly   Jan 1, 2010 - Dec 31, 2010   (Update)

Date	Open	High	Low	Close	Volume
Dec 31, 2010	43.87	44.33	43.72	44.17	4,288,633
Dec 30, 2010	43.75	44.08	43.43	43.89	5,072,979
Dec 29, 2010	43.71	44.10	43.70	43.95	5,150,671
Dec 28, 2010	44.01	44.20	43.99	44.11	3,253,454
Dec 27, 2010	43.98	44.00	43.63	43.97	2,797,542
Dec 23, 2010	44.00	44.00	44.00	44.00	0
Dec 22, 2010	43.43	43.71	43.20	43.61	4,198,035
Dec 21, 2010	43.77	43.88	43.51	43.54	4,822,785

B **Export**

⬇ Download to spreadsheet

Figure A.13 Daily historical prices of BP stock for 2010. Clicking on "Download to spreadsheet" will allow the user to download the historical prices in a Comma Separated Values format.

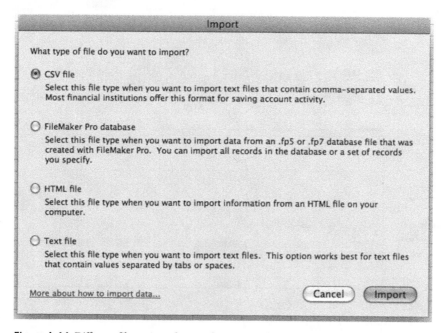

Figure A.14 Different file options that can be imported using Excel. In case of the BP stock prices, select CSV file.

Text Import Wizard – Step 1 of 3

The Text Wizard has determined that your data is Delimited.

If this is correct, choose Next, or choose the Data Type that best describes your data.

Original data type

Choose the file type that best describes your data:

- ⦿ Delimited – Characters such as commas or tabs separate each field.
- ○ Fixed width – Fields are aligned in columns with spaces between each field.

Start import at row: 1   File origin: Macintosh

Data preview

Preview of file Shiva:Users:pramodabichandani:Downloads:data.csv.

```
1 0°øDate,Open,High,Low,Close,Volume
2 31-Dec-10,43.87,44.33,43.72,44.17,4288633
3 30-Dec-10,43.75,44.08,43.43,43.89,5072979
4 29-Dec-10,43.71,44.10,43.70,43.95,5150671
5 28-Dec-10,44.01,44.20,43.99,44.11,3253454
6 27-Dec-10,43.98,44.00,43.63,43.97,2797542
```

Cancel   < Back   Next >   Finish

Figure A.15 The BP stock price file is a CSV file, and hence the data in this file are delimited using commas. The Delimited option is selected to reflect this.

In step 2, a comma is chosen as the delimiter (Fig. A.16).

The final step (Fig. A.17) allows you to select the data format for each column. In this case, the columns are already in a usable format. Simply press finish to complete the import text procedure.

As an additional step, you can select the starting cell location where the imported data should be placed in the Excel spreadsheet. The starting cell can be selected by clicking on it. Alternatively, you can import this data in a new spreadsheet in the Excel Workbook by selecting the New sheet option (Fig. A.18). Now the data will be displayed on the spreadsheet. Save this Excel Workbook as BP NYSE2010.xlsx.

3.3 Representing Data Clearly

It is critical to set up the imported data in a visually coherent manner to avoid obfuscation of data. If correctly imported, the spreadsheet in the BP NYSE2010.xlsx workbook should appear similar to that shown in Figure A.19. The first row of this spreadsheet is called the Header Row because it houses the names of each of the columns. Column A, for example, is titled Date since the dates when BP stock prices were

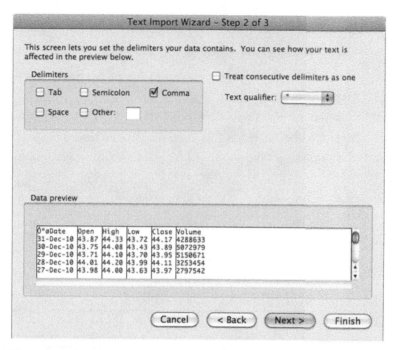

Figure A.16 Step 2 of the Text Import Wizard requires the user to select the particular delimiter being used in the file. In case of the BP stock prices, a comma is selected because the file is a .csv file.

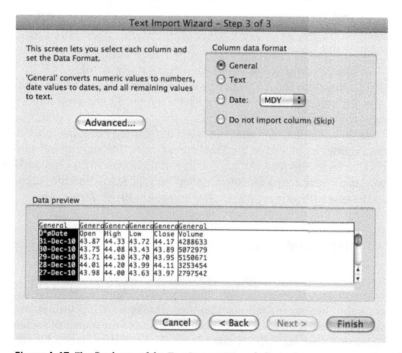

Figure A.17 The final step of the Text Import Wizard allows the user to enter the column data format. Since the columns are already in a usable format for the BP stock prices, we press Finish.

Figure A.18 Click on a cell in the Excel spreadsheet to indicate the starting cell location. In this case we select the first cell, A1, of the spreadsheet.

	A	B	C	D	E	F
1	Date	Open	High	Low	Close	Volume
2	31-Dec-10	43.87	44.33	43.72	44.17	4288633
3	30-Dec-10	43.75	44.08	43.43	43.89	5072979
4	29-Dec-10	43.71	44.1	43.7	43.95	5150671
5	28-Dec-10	44.01	44.2	43.99	44.11	3253454
6	27-Dec-10	43.98	44	43.63	43.97	2797542
7	23-Dec-10	44	44	44	44	0
8	22-Dec-10	43.43	43.71	43.2	43.61	4198035
9	21-Dec-10	43.77	43.88	43.51	43.54	4822785
10	20-Dec-10	43.75	43.75	43.2	43.68	7615564
11	17-Dec-10	43.58	43.76	43.16	43.25	6516467
12	16-Dec-10	43.89	44.02	43.59	43.75	10269985
13	15-Dec-10	44.4	44.74	43.29	43.86	16958988
14	14-Dec-10	44.83	44.83	44.06	44.44	17296056
15	13-Dec-10	43.48	43.8	43.41	43.43	6414485
16	10-Dec-10	42.91	43.35	42.89	43.24	6617173
17	9-Dec-10	43.16	43.34	42.57	42.79	7977341
18	8-Dec-10	43.59	43.61	42.87	43.27	6567300
19	7-Dec-10	43.39	43.5	42.85	42.89	8460564
20	6-Dec-10	42.08	42.92	42.05	42.81	9966528
21	3-Dec-10	41.46	41.63	41.04	41.49	8019874
22	2-Dec-10	40.78	41.42	40.77	41.32	10236958
23	1-Dec-10	40.55	40.73	40.15	40.62	8716107

Figure A.19 BP stock performance NYSE 2010 imported from a .csv file.

recorded are stored in the cells of this column. The other columns of the spreadsheet provide specific information about the stock performance. These are explained in Table A.1.

Table A.1 Description of the Header Column Fields

Column Name	Specific Stock Information Provided
Date	Date on which the stock prices were recorded
Open	The price at which a stock first trades when the NYSE begins trading on a given date
High	The highest price that the stock trades at on a given date
Low	The lowest price that the stock trades at on a given date
Close	The final price at which the stock is traded on a given date
Volume	The total number of stocks traded on a given date

The data can be sorted from Jan. 1, 2010, to Dec. 31, 2010, by using Excel's sort functionalities (Figs. A.20 and A.21). Using the Sort option available in the Data menu, you can sort data in a variety of ways. In this case, the data can be arranged in chronological order.

Using Excel's standard formatting features such as font sizes, font colors, insert rows and columns in a spreadsheet, cell fill colors, and merge cells, you can create a visually effective representation of the data stored in the file (Fig. A.22). As you can see, a title is added in the file by inserting a row at the beginning of the sheet, selecting a few cells, and merging them. Color-coding allows you to clearly identify different types of cells. Students are encouraged to explore different formatting features of Excel.

3.4 Performing Basic Math

Excel can be used as a calculator to perform basic mathematical operations, including addition, subtraction, division, and multiplication. Additionally, trigonometric functions can be used. An = sign is entered in a cell before entering a mathematical formula in it. Figure A.23 depicts the use of the Excel to add two numbers.

Excel can be used to ascertain the product of two numbers, as shown in Figure A.24.

Similarly, subtractions and divisions can be performed as shown in Figures A.25 and A.26, respectively.

Excel provides inbuilt functionality to perform various mathematical operations. For example, the inbuilt function sum() can be used to calculate the sum of various cells, as shown in Figure A.27.

Similarly, Excel function product() can be used to calculate the product of several numbers. Figure A.28 shows the use of this function to calculate the product of two numbers.

Figure A.20 Since the columns have names, select the option "My list has headers."

Figure A.21 Excel's Sort dialog box can help users sort the data in various ways.

	A	B	C	D	E	F
1	**BP NYSE STOCK PERFORMANCE 2010**					
2	Date	Open	High	Low	Close	Volume
3	1-Jan-10	57.97	57.97	57.97	57.97	0
4	4-Jan-10	59.3	59.45	59.08	59.15	3956568
5	5-Jan-10	59.65	59.9	59.31	59.57	4109563
6	6-Jan-10	59.52	59.92	59.34	59.88	6230175
7	7-Jan-10	59.92	60	59.69	59.86	4432171
8	8-Jan-10	59.79	60.06	59.67	60	3786118
9	11-Jan-10	61.54	61.97	61.45	61.89	8229519
10	12-Jan-10	60.86	61.69	60.8	61.5	6127216
11	13-Jan-10	61.31	61.98	61	61.8	5688621
12	14-Jan-10	61.22	61.75	61.16	61.73	3330668
13	15-Jan-10	61.64	61.64	61.64	61.64	0
14	18-Jan-10	61.64	61.64	61.64	61.64	0

Figure A.22 The imported data are sorted in a chronological order using Excel's sort options. An additional row is added and the cells are merged to provide a title for the dataset.

Figure A.23 Excel can be used to perform basic mathematical operations like addition. Here, two numbers are being added. An = sign has to be entered before entering any mathematical formula into an Excel cell.

Figure A.24 Excel being used to perform multiplication of two numbers.

Figure A.25 Excel being used to perform subtraction of two numbers.

Figure A.26 Excel being used to perform division of two numbers.

Figure A.27 Excel being used to perform multiplication of two numbers.

Figure A.28 Excel being used to perform multiplication of two numbers.

3.5   Using Basic Descriptive Statistics

Once you have imported the data and visually arranged it, the next step is to analyze it using statistical techniques. There are two main types of statistical techniques that we will use: descriptive and predictive techniques.

Descriptive statistics, as the name suggests, summarizes several properties of the data at hand. These properties include sample size, minimum value, maximum value, range, mean, median, mode, standard deviation, variance, etc. Predictive statistics refers to statistical methods that allow you to generate predictions by analyzing current and historical observations. In the following discussions, we will define these quantities mathematically and use Excel to ascertain their values.

While working with data, it is important to identify whether the data at hand are (1) a sample of a larger dataset or (2) the complete set of data required for the analyses. Accordingly, two types of datasets are possible:

1. Population dataset: This dataset contains all possible measurements that are of interest in a particular study.
2. Sample dataset: This dataset contains only a portion of the population dataset. The sample dataset is representative of the population dataset from which it was selected.

For example, in case of BP stock prices, the file BP NYSE2010.xlsx contains stock prices for 2010 only. For the purpose of our analyses, we are only interested in the BP stock behavior for 2010. The file BP NYSE2010.xlsx contains all the data required to analyze its behavior for 2010. Thus, in this respect, the file BP NYSE2010.xlsx is a population dataset for 2010.

However, if we were to look at the performance of BP stock for the years 2000 to 2012, the file BP NYSE2010.xlsx would be considered a sample dataset since it contains information about only one year (2010), whereas the study at hand involves data ranging from 2000 to 2012.

In Table A.2, four basic descriptive statistics are explained along with the Excel function required to implement them. These are sample size, minimum value, maximum value, and range.

Table A.2 Basic Descriptive Statistics

Statistic	Description	Excel Function
Sample Size	Total number of data points	count(startCell:endCell)
Minimum Value	Minimum value of the stock prices	min(startCell:endCell)
Maximum Value	Maximum value of the stock prices	max(startCell:endCell)
Range	Range of values of the stock prices	range = max(startCell:endCell)- min(startCell:endCell)

On examining the closing price of BP stock in 2010, we find that the stock was traded for 259 days, as shown in Figures A.29 and A.30.

The minimum closing price was $27.02 and the maximum closing price was $62.32, thus providing a range of $35.3 (Figs. A.31, A.32, and A.33, respectively).

	A	B	C	D	E	F	G	H	I	J
1	**BP NYSE STOCK PERFORMANCE 2010**									
2	Date	Open	High	Low	Close	Volume				
3	1-Jan-10	57.97	57.97	57.97	57.97	0				
4	4-Jan-10	59.3	59.45	59.08	59.15	3956568		Descriptive Statistics	Value	
5	5-Jan-10	59.65	59.9	59.31	59.57	4109563		Number of Data Points	=count(E3:E261)	
6	6-Jan-10	59.52	59.92	59.34	59.88	6230175		Minimum Closing Price		
7	7-Jan-10	59.92	60	59.69	59.86	4432171		Maximum Closing Price		
8	8-Jan-10	59.79	60.06	59.67	60	3786118		Range (Max Closing price - Min Closing Price)		
9	11-Jan-10	61.54	61.97	61.45	61.89	8229519				
10	12-Jan-10	60.86	61.69	60.8	61.5	6127216				
11	13-Jan-10	61.31	61.98	61	61.8	5688621				
12	14-Jan-10	61.22	61.75	61.16	61.73	3330668				
13	15-Jan-10	61.64	61.64	61.64	61.64	0				

Figure A.29 Selecting the range of closing stock prices to calculate the total number of data points.

A

	A	B	C	D	E	F	G	H	I	J
1	**BP NYSE STOCK PERFORMANCE 2010**									
2	Date	Open	High	Low	Close	Volume				
3	1-Jan-10	57.97	57.97	57.97	57.97	0				
4	4-Jan-10	59.3	59.45	59.08	59.15	3956568		Descriptive Statistics	Value	
5	5-Jan-10	59.65	59.9	59.31	59.57	4109563		Number of Data Points	=count(E3:E261)	
6	6-Jan-10	59.52	59.92	59.34	59.88	6230175		Minimum Closing Price		
7	7-Jan-10	59.92	60	59.69	59.86	4432171		Maximum Closing Price		
8	8-Jan-10	59.79	60.06	59.67	60	3786118		Range (Max Closing price - Min Closing Price)		
9	11-Jan-10	61.54	61.97	61.45	61.89	8229519				
10	12-Jan-10	60.86	61.69	60.8	61.5	6127216				
11	13-Jan-10	61.31	61.98	61	61.8	5688621				
12	14-Jan-10	61.22	61.75	61.16	61.73	3330668				
13	15-Jan-10	61.64	61.64	61.64	61.64	0				

B

	A	B	C	D	E	F	G	H	I
1	**BP NYSE STOCK PERFORMANCE 2010**								
2	Date	Open	High	Low	Close	Volume			
3	1-Jan-10	57.97	57.97	57.97	57.97	0			
4	4-Jan-10	59.3	59.45	59.08	59.15	3956568		Descriptive Statistics	Value
5	5-Jan-10	59.65	59.9	59.31	59.57	4109563		Number of Data Points	259
6	6-Jan-10	59.52	59.92	59.34	59.88	6230175		Minimum Closing Price	
7	7-Jan-10	59.92	60	59.69	59.86	4432171		Maximum Closing Price	
8	8-Jan-10	59.79	60.06	59.67	60	3786118		Range (Max Closing price - Min Closing Price)	
9	11-Jan-10	61.54	61.97	61.45	61.89	8229519			
10	12-Jan-10	60.86	61.69	60.8	61.5	6127216			
11	13-Jan-10	61.31	61.98	61	61.8	5688621			
12	14-Jan-10	61.22	61.75	61.16	61.73	3330668			
13	15-Jan-10	61.64	61.64	61.64	61.64	0			

Figure A.30 Use of count() function to calculate the number of samples in BP NYSE2010.xlsx.

3.5.1 Measures of the Central Tendency

Let $S_1 \ldots S_n$ denote the BP closing stock prices stored in BP NYSE2010.xlsx, where $n = 259$. The central tendency is the tendency of the stock prices to cluster around some value. There are several measures of central tendency. The three main measures are the mean, median, and mode.

A

	A	B	C	D	E	F	G	H	I
1	BP NYSE STOCK PERFORMANCE 2010								
2	Date	Open	High	Low	Close	Volume			
3	1-Jan-10	57.97	57.97	57.97	57.97	0			
4	4-Jan-10	59.3	59.45	59.08	59.15	3956568		Descriptive Statistics	Value
5	5-Jan-10	59.65	59.9	59.31	59.57	4109563		Number of Data Points	259
6	6-Jan-10	59.52	59.92	59.34	59.88	6230175		Minimum Closing Price	=min(E3:E261)
7	7-Jan-10	59.92	60	59.69	59.86	4432171		Maximum Closing Price	
8	8-Jan-10	59.79	60.06	59.67	60	3786118		Range (Max Closing price - Min Closing Price)	
9	11-Jan-10	61.54	61.97	61.45	61.89	8229519			
10	12-Jan-10	60.86	61.69	60.8	61.5	6127216			
11	13-Jan-10	61.31	61.98	61	61.8	5688621			
12	14-Jan-10	61.22	61.75	61.16	61.73	3330668			
13	15-Jan-10	61.64	61.64	61.64	61.64	0			

B

	A	B	C	D	E	F	G	H	I
1	BP NYSE STOCK PERFORMANCE 2010								
2	Date	Open	High	Low	Close	Volume			
3	1-Jan-10	57.97	57.97	57.97	57.97	0			
4	4-Jan-10	59.3	59.45	59.08	59.15	3956568		Descriptive Statistics	Value
5	5-Jan-10	59.65	59.9	59.31	59.57	4109563		Number of Data Points	259
6	6-Jan-10	59.52	59.92	59.34	59.88	6230175		Minimum Closing Price	27.02
7	7-Jan-10	59.92	60	59.69	59.86	4432171		Maximum Closing Price	
8	8-Jan-10	59.79	60.06	59.67	60	3786118		Range (Max Closing price - Min Closing Price)	
9	11-Jan-10	61.54	61.97	61.45	61.89	8229519			
10	12-Jan-10	60.86	61.69	60.8	61.5	6127216			
11	13-Jan-10	61.31	61.98	61	61.8	5688621			
12	14-Jan-10	61.22	61.75	61.16	61.73	3330668			
13	15-Jan-10	61.64	61.64	61.64	61.64	0			

Figure A.31 Use of min() function to calculate the minimum closing price of BP stock in 2010.

A

	A	B	C	D	E	F	G	H	I
1	BP NYSE STOCK PERFORMANCE 2010								
2	Date	Open	High	Low	Close	Volume			
3	1-Jan-10	57.97	57.97	57.97	57.97	0			
4	4-Jan-10	59.3	59.45	59.08	59.15	3956568		Descriptive Statistics	Value
5	5-Jan-10	59.65	59.9	59.31	59.57	4109563		Number of Data Points	259
6	6-Jan-10	59.52	59.92	59.34	59.88	6230175		Minimum Closing Price	27.02
7	7-Jan-10	59.92	60	59.69	59.86	4432171		Maximum Closing Price	=MAX(E3:E261)
8	8-Jan-10	59.79	60.06	59.67	60	3786118		Range (Max Closing price - Min Closing Price)	
9	11-Jan-10	61.54	61.97	61.45	61.89	8229519			
10	12-Jan-10	60.86	61.69	60.8	61.5	6127216			
11	13-Jan-10	61.31	61.98	61	61.8	5688621			
12	14-Jan-10	61.22	61.75	61.16	61.73	3330668			
13	15-Jan-10	61.64	61.64	61.64	61.64	0			

B

	A	B	C	D	E	F	G	H	I
1	BP NYSE STOCK PERFORMANCE 2010								
2	Date	Open	High	Low	Close	Volume			
3	1-Jan-10	57.97	57.97	57.97	57.97	0			
4	4-Jan-10	59.3	59.45	59.08	59.15	3956568		Descriptive Statistics	Value
5	5-Jan-10	59.65	59.9	59.31	59.57	4109563		Number of Data Points	259
6	6-Jan-10	59.52	59.92	59.34	59.88	6230175		Minimum Closing Price	27.02
7	7-Jan-10	59.92	60	59.69	59.86	4432171		Maximum Closing Price	62.32
8	8-Jan-10	59.79	60.06	59.67	60	3786118		Range (Max Closing price - Min Closing Price)	
9	11-Jan-10	61.54	61.97	61.45	61.89	8229519			
10	12-Jan-10	60.86	61.69	60.8	61.5	6127216			
11	13-Jan-10	61.31	61.98	61	61.8	5688621			
12	14-Jan-10	61.22	61.75	61.16	61.73	3330668			
13	15-Jan-10	61.64	61.64	61.64	61.64	0			

Figure A.32 Use of max() function to calculate the maximum closing price of BP stock in 2010.

A

	A	B	C	D	E	F	G	H	I
1	**BP NYSE STOCK PERFORMANCE 2010**								
2	Date	Open	High	Low	Close	Volume			
3	1-Jan-10	57.97	57.97	57.97	57.97	0			
4	4-Jan-10	59.3	59.45	59.08	59.15	3956568		Descriptive Statistics	Value
5	5-Jan-10	59.65	59.9	59.31	59.57	4109563		Number of Data Points	259
6	6-Jan-10	59.52	59.92	59.34	59.88	6230175		Minimum Closing Price	27.02
7	7-Jan-10	59.92	60	59.69	59.86	4432171		Maximum Closing Price	62.32
8	8-Jan-10	59.79	60.06	59.67	60	3786118		Range (Max Closing price - Min Closing Price)	=I7-I6
9	11-Jan-10	61.54	61.97	61.45	61.89	8229519			
10	12-Jan-10	60.86	61.69	60.8	61.5	6127216			
11	13-Jan-10	61.31	61.98	61	61.8	5688621			
12	14-Jan-10	61.22	61.75	61.16	61.73	3330668			
13	15-Jan-10	61.64	61.64	61.64	61.64	0			

B

	A	B	C	D	E	F	G	H	I
1	**BP NYSE STOCK PERFORMANCE 2010**								
2	Date	Open	High	Low	Close	Volume			
3	1-Jan-10	57.97	57.97	57.97	57.97	0			
4	4-Jan-10	59.3	59.45	59.08	59.15	3956568		Descriptive Statistics	Value
5	5-Jan-10	59.65	59.9	59.31	59.57	4109563		Number of Data Points	259
6	6-Jan-10	59.52	59.92	59.34	59.88	6230175		Minimum Closing Price	27.02
7	7-Jan-10	59.92	60	59.69	59.86	4432171		Maximum Closing Price	62.32
8	8-Jan-10	59.79	60.06	59.67	60	3786118		Range (Max Closing price - Min Closing Price)	35.3
9	11-Jan-10	61.54	61.97	61.45	61.89	8229519			
10	12-Jan-10	60.86	61.69	60.8	61.5	6127216			
11	13-Jan-10	61.31	61.98	61	61.8	5688621			
12	14-Jan-10	61.22	61.75	61.16	61.73	3330668			
13	15-Jan-10	61.64	61.64	61.64	61.64	0			

Figure A.33 Calculating the range of closing BP stock prices for NYSE in 2010.

1. The *mean* closing stock price of the BP stock, S_{avg}, is the arithmetic mean of the daily closing stock prices stored in BP NYSE2010.xlsx:

$$S_{avg} = (S_1 + \ldots + S_n)/n \tag{1}$$

2. The *median* closing stock price of the BP stock is the numerical value that separates the higher half of the BP closing stock prices from the lower half.
3. The *mode* of the closing stock price of the BP stock is the closing price value that appears the most throughout 2010.

Excel can be used to calculate all three of these measures of central tendency.

Figure A.34 depicts the use of Excel function sum() to calculate the arithmetic mean defined in (1). Instead of using the (1), one can use an inbuilt Excel command, average(), to calculate the arithmetic means as well. The use of this function is displayed in Figure A.35.

Figures A.36 and A.37 depict the use of median() and mode.sngl() functions, respectively. These functions are used to calculate the median and mode of the BP closing stock prices, respectively.

3.5.2 Measures of Spread

While the measures of central tendency describe the tendency of the data to cluster around a certain value, the measures of spread describe the spread or variability of the data. There are several measures of spread. The two most common ones are standard

A

	A	B	C	D	E	F	G	H	I
1	BP NYSE STOCK PERFORMANCE 2010								
2	Date	Open	High	Low	Close	Volume			
3	1-Jan-10	57.97	57.97	57.97	57.97	0			
4	4-Jan-10	59.3	59.45	59.08	59.15	3956568		Descriptive Statistics	Value
5	5-Jan-10	59.65	59.9	59.31	59.57	4109563		Number of Data Points	259
6	6-Jan-10	59.52	59.92	59.34	59.88	6230175		Minimum Closing Price	27.02
7	7-Jan-10	59.92	60	59.69	59.86	4432171		Maximum Closing Price	62.32
8	8-Jan-10	59.79	60.06	59.67	60	3786118		Range (Max Closing price - Min Closing Price)	35.3
9	11-Jan-10	61.54	61.97	61.45	61.89	8229519			
10	12-Jan-10	60.86	61.69	60.8	61.5	6127216		Mean (using formula)	=sum(E3:E261)/IS
11	13-Jan-10	61.31	61.98	61	61.8	5688621			
12	14-Jan-10	61.22	61.75	61.16	61.73	3330668			
13	15-Jan-10	61.64	61.64	61.64	61.64	0			

B

	A	B	C	D	E	F	G	H	I
1	BP NYSE STOCK PERFORMANCE 2010								
2	Date	Open	High	Low	Close	Volume			
3	1-Jan-10	57.97	57.97	57.97	57.97	0			
4	4-Jan-10	59.3	59.45	59.08	59.15	3956568		Descriptive Statistics	Value
5	5-Jan-10	59.65	59.9	59.31	59.57	4109563		Number of Data Points	259
6	6-Jan-10	59.52	59.92	59.34	59.88	6230175		Minimum Closing Price	27.02
7	7-Jan-10	59.92	60	59.69	59.86	4432171		Maximum Closing Price	62.32
8	8-Jan-10	59.79	60.06	59.67	60	3786118		Range (Max Closing price - Min Closing Price)	35.3
9	11-Jan-10	61.54	61.97	61.45	61.89	8229519			
10	12-Jan-10	60.86	61.69	60.8	61.5	6127216		Mean (using formula)	45.39540541
11	13-Jan-10	61.31	61.98	61	61.8	5688621			
12	14-Jan-10	61.22	61.75	61.16	61.73	3330668			
13	15-Jan-10	61.64	61.64	61.64	61.64	0			

Figure A.34 Using Equation (1) to calculate the arithmetic mean of the closing stock prices of BP in 2010.

A

	A	B	C	D	E	F	G	H	I
1	BP NYSE STOCK PERFORMANCE 2010								
2	Date	Open	High	Low	Close	Volume			
3	1-Jan-10	57.97	57.97	57.97	57.97	0			
4	4-Jan-10	59.3	59.45	59.08	59.15	3956568		Descriptive Statistics	Value
5	5-Jan-10	59.65	59.9	59.31	59.57	4109563		Number of Data Points	259
6	6-Jan-10	59.52	59.92	59.34	59.88	6230175		Minimum Closing Price	27.02
7	7-Jan-10	59.92	60	59.69	59.86	4432171		Maximum Closing Price	62.32
8	8-Jan-10	59.79	60.06	59.67	60	3786118		Range (Max Closing price - Min Closing Price)	35.3
9	11-Jan-10	61.54	61.97	61.45	61.89	8229519			
10	12-Jan-10	60.86	61.69	60.8	61.5	6127216		Mean (using formula)	45.39540541
11	13-Jan-10	61.31	61.98	61	61.8	5688621		Mean (using Excel Function)	=AVERAGE(E3:E261)
12	14-Jan-10	61.22	61.75	61.16	61.73	3330668			
13	15-Jan-10	61.64	61.64	61.64	61.64	0			

B

	A	B	C	D	E	F	G	H	I
1	BP NYSE STOCK PERFORMANCE 2010								
2	Date	Open	High	Low	Close	Volume			
3	1-Jan-10	57.97	57.97	57.97	57.97	0			
4	4-Jan-10	59.3	59.45	59.08	59.15	3956568		Descriptive Statistics	Value
5	5-Jan-10	59.65	59.9	59.31	59.57	4109563		Number of Data Points	259
6	6-Jan-10	59.52	59.92	59.34	59.88	6230175		Minimum Closing Price	27.02
7	7-Jan-10	59.92	60	59.69	59.86	4432171		Maximum Closing Price	62.32
8	8-Jan-10	59.79	60.06	59.67	60	3786118		Range (Max Closing price - Min Closing Price)	35.3
9	11-Jan-10	61.54	61.97	61.45	61.89	8229519			
10	12-Jan-10	60.86	61.69	60.8	61.5	6127216		Mean (using formula)	45.39540541
11	13-Jan-10	61.31	61.98	61	61.8	5688621		Mean (using Excel Function)	45.39540541
12	14-Jan-10	61.22	61.75	61.16	61.73	3330668			
13	15-Jan-10	61.64	61.64	61.64	61.64	0			

Figure A.35 The mean closing price of BP stock in 2010 was $45.39.

A

BP NYSE STOCK PERFORMANCE 2010

Date	Open	High	Low	Close	Volume
1-Jan-10	57.97	57.97	57.97	57.97	0
4-Jan-10	59.3	59.45	59.08	59.15	3956568
5-Jan-10	59.65	59.9	59.31	59.57	4109563
6-Jan-10	59.52	59.92	59.34	59.88	6230175
7-Jan-10	59.92	60	59.69	59.86	4432171
8-Jan-10	59.79	60.06	59.67	60	3786118
11-Jan-10	61.54	61.97	61.45	61.89	8229519
12-Jan-10	60.86	61.69	60.8	61.5	6127216
13-Jan-10	61.31	61.98	61	61.8	5688621
14-Jan-10	61.22	61.75	61.16	61.73	3330668
15-Jan-10	61.64	61.64	61.64	61.64	0
18-Jan-10	61.64	61.64	61.64	61.64	0

Descriptive Statistics	Value
Number of Data Points	259
Minimum Closing Price	27.02
Maximum Closing Price	62.32
Range (Max Closing price - Min Closing Price)	35.3
Mean (using formula)	45.39540541
Mean (using Excel Function)	45.39540541
Median	=median(E3:E261)

B

BP NYSE STOCK PERFORMANCE 2010

Date	Open	High	Low	Close	Volume
1-Jan-10	57.97	57.97	57.97	57.97	0
4-Jan-10	59.3	59.45	59.08	59.15	3956568
5-Jan-10	59.65	59.9	59.31	59.57	4109563
6-Jan-10	59.52	59.92	59.34	59.88	6230175
7-Jan-10	59.92	60	59.69	59.86	4432171
8-Jan-10	59.79	60.06	59.67	60	3786118
11-Jan-10	61.54	61.97	61.45	61.89	8229519
12-Jan-10	60.86	61.69	60.8	61.5	6127216
13-Jan-10	61.31	61.98	61	61.8	5688621
14-Jan-10	61.22	61.75	61.16	61.73	3330668
15-Jan-10	61.64	61.64	61.64	61.64	0
18-Jan-10	61.64	61.64	61.64	61.64	0

Descriptive Statistics	Value
Number of Data Points	259
Minimum Closing Price	27.02
Maximum Closing Price	62.32
Range (Max Closing price - Min Closing Price)	35.3
Mean (using formula)	45.39540541
Mean (using Excel Function)	45.39540541
Median	42.81

Figure A.36 Use of the median() function to calculate the median closing price of BP stock in 2010.

A

BP NYSE STOCK PERFORMANCE 2010

Date	Open	High	Low	Close	Volume
1-Jan-10	57.97	57.97	57.97	57.97	0
4-Jan-10	59.3	59.45	59.08	59.15	3956568
5-Jan-10	59.65	59.9	59.31	59.57	4109563
6-Jan-10	59.52	59.92	59.34	59.88	6230175
7-Jan-10	59.92	60	59.69	59.86	4432171
8-Jan-10	59.79	60.06	59.67	60	3786118
11-Jan-10	61.54	61.97	61.45	61.89	8229519
12-Jan-10	60.86	61.69	60.8	61.5	6127216
13-Jan-10	61.31	61.98	61	61.8	5688621
14-Jan-10	61.22	61.75	61.16	61.73	3330668
15-Jan-10	61.64	61.64	61.64	61.64	0
18-Jan-10	61.64	61.64	61.64	61.64	0

Descriptive Statistics	Value
Number of Data Points	259
Minimum Closing Price	27.02
Maximum Closing Price	62.32
Range (Max Closing price - Min Closing Price)	35.3
Mean (using formula)	45.39540541
Mean (using Excel Function)	45.39540541
Median	42.81
Mode	=MODE.SNGL(E3:E261)

B

BP NYSE STOCK PERFORMANCE 2010

Date	Open	High	Low	Close	Volume
1-Jan-10	57.97	57.97	57.97	57.97	0
4-Jan-10	59.3	59.45	59.08	59.15	3956568
5-Jan-10	59.65	59.9	59.31	59.57	4109563
6-Jan-10	59.52	59.92	59.34	59.88	6230175
7-Jan-10	59.92	60	59.69	59.86	4432171
8-Jan-10	59.79	60.06	59.67	60	3786118
11-Jan-10	61.54	61.97	61.45	61.89	8229519
12-Jan-10	60.86	61.69	60.8	61.5	6127216
13-Jan-10	61.31	61.98	61	61.8	5688621
14-Jan-10	61.22	61.75	61.16	61.73	3330668
15-Jan-10	61.64	61.64	61.64	61.64	0
18-Jan-10	61.64	61.64	61.64	61.64	0

Descriptive Statistics	Value
Number of Data Points	259
Minimum Closing Price	27.02
Maximum Closing Price	62.32
Range (Max Closing price - Min Closing Price)	35.3
Mean (using formula)	45.39540541
Mean (using Excel Function)	45.39540541
Median	42.81
Mode	59.88

Figure A.37 Use of the mode.sngl() function to calculate the mode of the closing prices of BP stock in 2010.

There are 10 bins in this histogram. The bin minimum and maximum limits are listed in Table A.3. The frequency is listed next to each bin. The frequency is defined as the number of data points that belong to each bin. When correctly designed, the total frequency of a histogram should be equal to the number of data points in a dataset. In this case, the total frequency should equal 259.

Table A.3 Bin Information for the Histogram Depicting BP Closing Stock Prices in 2010.

Bin Minimum	Bin Maximum	Frequency
27	30.6	12
30.7	34.3	11
34.4	38	31
38.1	41.7	65
41.8	45.4	43
45.5	49.1	7
49.2	52.8	7
52.9	56.5	28
56.6	60.2	45
60.3	63.9	10
	Total	259

Selecting the correct number of bins and bin size is critical for creating a meaningful histogram. While several approaches to ascertaining the optimal number of bins in a histogram have been developed, more often than not we make an informed decision about the number of bins based on our knowledge of the data and the underlying process that generates it.

As far as the bin size is concerned, a good starting point is to divide the entire data range equally into the number of bins. If n_{bin} is the number of bins, the bin size can be defined as

$$\text{bin size} = \text{range}/n_{bin} \tag{3}$$

Excel can create a histogram in several ways. We will create a histogram by first enumerating the bins and ascertaining the frequencies in each bin. To create a histogram with 10 bins, we implement Equation 3 in Excel, as shown in Figures A.41 and A.42.

It is clear that the bin size for the BP closing stock prices in 2010 should be 3.53. We generate the bin minimum and maximum limits as shown in Figures A.43, A.44, and A.45. To create the minimum limit of the first bin, we round down the minimum value of the stock closing price to the nearest integer using the Excel function rounddown().

Next, the upper limit is determined by adding the bin size to the minimum limit. The minimum limit of the next bin is found by simply incrementing the previous bin maximum limit by a very small amount. In this case, we add 0.01 to previous bin maximum limit, as shown in Figure A.45. Repeating this process allows us to create the 10 bins with their minimum and maximum ranges. Using Excel's frequency() function, we can

Descriptive Statistics	Value (in Dollars)
Number of Data Points	259
Minimum Closing Price	27.02
Maximum Closing Price	62.32
Range (Max Closing price - Min Closing Price)	35.3
Average Closing Price	45.4
Standard Deviation of Closing Prices	9.391236824
Variance of the Closing Prices	88.19532908
number of bins	10
Bin size	=J8/J13

Figure A.41 Bin size determination—divide the range by the desired number of bins.

A

	A	B	C	D	E	F		H	I	J
1	**BP NYSE STOCK PERFORMANCE 2010**									
2	Date	Open	High	Low	Close	Volume				
3	1-Jan-10	57.97	57.97	57.97	57.97	0				
4	4-Jan-10	59.3	59.45	59.08	59.15	3956568		Descriptive Statistics		Value (in Dollars)
5	5-Jan-10	59.65	59.9	59.31	59.57	4109563		Number of Data Points		259
6	6-Jan-10	59.52	59.92	59.34	59.88	6230175		Minimum Closing Price		27.02
7	7-Jan-10	59.92	60	59.69	59.86	4432171		Maximum Closing Price		62.32
8	8-Jan-10	59.79	60.06	59.67	60	3786118		Range (Max Closing price - Min Closing Price)		35.3
9	11-Jan-10	61.54	61.97	61.45	61.89	8229519		Average Closing Price		45.4
10	12-Jan-10	60.86	61.69	60.8	61.5	6127216		Standard Deviation of Closing Prices		9.391236824
11	13-Jan-10	61.31	61.98	61	61.8	5688621		Variance of the Closing Prices		88.19532908
12	14-Jan-10	61.22	61.75	61.16	61.73	3330668				
13	15-Jan-10	61.64	61.64	61.64	61.64	0		number of bins		10
14	18-Jan-10	61.64	61.64	61.64	61.64	0		Bin size		=J8/J13

B

	A	B	C	D	E	F		H	I	J
1	**BP NYSE STOCK PERFORMANCE 2010**									
2	Date	Open	High	Low	Close	Volume				
3	1-Jan-10	57.97	57.97	57.97	57.97	0				
4	4-Jan-10	59.3	59.45	59.08	59.15	3956568		Descriptive Statistics		Value (in Dollars)
5	5-Jan-10	59.65	59.9	59.31	59.57	4109563		Number of Data Points		259
6	6-Jan-10	59.52	59.92	59.34	59.88	6230175		Minimum Closing Price		27.02
7	7-Jan-10	59.92	60	59.69	59.86	4432171		Maximum Closing Price		62.32
8	8-Jan-10	59.79	60.06	59.67	60	3786118		Range (Max Closing price - Min Closing Price)		35.3
9	11-Jan-10	61.54	61.97	61.45	61.89	8229519		Average Closing Price		45.4
10	12-Jan-10	60.86	61.69	60.8	61.5	6127216		Standard Deviation of Closing Prices		9.391236824
11	13-Jan-10	61.31	61.98	61	61.8	5688621		Variance of the Closing Prices		88.19532908
12	14-Jan-10	61.22	61.75	61.16	61.73	3330668				
13	15-Jan-10	61.64	61.64	61.64	61.64	0		number of bins		10
14	18-Jan-10	61.64	61.64	61.64	61.64	0		Bin size		3.53
15	19-Jan-10	61.76	62.38	61.76	62.32	4914423				

Figure A.42 Determining bin sizes using Equation (3) in Excel for BP closing stock prices in 2010.

ascertain the number of closing stock prices that belong to a particular bin. The use of this function is depicted in Figure A.46.

Students are encouraged to read the help documentation associated with the frequency() function in Excel for more details.

The final step in creating the histogram is to plot the column graph. We use Excel's 2-D clustered column chart for this. On selecting this option from Excel's chart offerings, a blank

A

BP NYSE STOCK PERFORMANCE 2010

Date	Open	High	Low	Close	Volume
1-Jan-10	57.97	57.97	57.97	57.97	0
4-Jan-10	59.3	59.45	59.08	59.15	3956568
5-Jan-10	59.65	59.9	59.31	59.57	4109563
6-Jan-10	59.52	59.92	59.34	59.88	6230175
7-Jan-10	59.92	60	59.69	59.86	4432171
8-Jan-10	59.79	60.06	59.67	60	3786118
11-Jan-10	61.54	61.97	61.45	61.89	8229519
12-Jan-10	60.86	61.69	60.8	61.5	6127216
13-Jan-10	61.31	61.98	61	61.8	5688621
14-Jan-10	61.22	61.75	61.16	61.73	3330668
15-Jan-10	61.64	61.64	61.64	61.64	0
18-Jan-10	61.64	61.64	61.64	61.64	0
19-Jan-10	61.76	62.38	61.76	62.32	4914423
20-Jan-10	61.35	61.44	60.6	61.06	5410892
21-Jan-10	60.95	61.11	59.13	59.57	7184905
22-Jan-10	58.77	59.1	57.76	57.87	7297480
25-Jan-10	58.55	58.79	58.21	58.55	4041969
26-Jan-10	57.95	58.92	57.81	58.49	5685433

Descriptive Statistics	Value
Number of Data Points	259
Minimum Closing Price	27.02
Maximum Closing Price	62.32
Range (Max Closing price - Min Closing Price)	35.3
Average Closing Price	45.4
Standard Deviation of Closing Prices	9.391236824
Variance of the Closing Prices	88.19532908
Histogram	
number of bins	10
Bin size	3.53

Bin Minimum	Bin Maximum	Frequency
=ROUNDDOWN(J6,1)		

B

BP NYSE STOCK PERFORMANCE 2010

Date	Open	High	Low	Close	Volume
1-Jan-10	57.97	57.97	57.97	57.97	0
4-Jan-10	59.3	59.45	59.08	59.15	3956568
5-Jan-10	59.65	59.9	59.31	59.57	4109563
6-Jan-10	59.52	59.92	59.34	59.88	6230175
7-Jan-10	59.92	60	59.69	59.86	4432171
8-Jan-10	59.79	60.06	59.67	60	3786118
11-Jan-10	61.54	61.97	61.45	61.89	8229519
12-Jan-10	60.86	61.69	60.8	61.5	6127216
13-Jan-10	61.31	61.98	61	61.8	5688621
14-Jan-10	61.22	61.75	61.16	61.73	3330668
15-Jan-10	61.64	61.64	61.64	61.64	0
18-Jan-10	61.64	61.64	61.64	61.64	0
19-Jan-10	61.76	62.38	61.76	62.32	4914423
20-Jan-10	61.35	61.44	60.6	61.06	5410892
21-Jan-10	60.95	61.11	59.13	59.57	7184905
22-Jan-10	58.77	59.1	57.76	57.87	7297480
25-Jan-10	58.55	58.79	58.21	58.55	4041969
26-Jan-10	57.95	58.92	57.81	58.49	5685433
27-Jan-10	57.89	58.18	57.44	58.06	6026843

Descriptive Statistics	Value
Number of Data Points	259
Minimum Closing Price	27.02
Maximum Closing Price	62.32
Range (Max Closing price - Min Closing Price)	35.3
Average Closing Price	45.4
Standard Deviation of Closing Prices	9.391236824
Variance of the Closing Prices	88.19532908
Histogram	
number of bins	10
Bin size	3.53

Bin Minimum	Bin Maximum	Frequency
27		

Figure A.43 Generating the minimum limit of the first bin for the histogram shown in Figure. A.40. The Excel function rounddown() is used to round down the minimum value of the stock closing price to the nearest integer.

BP NYSE STOCK PERFORMANCE 2010

Date	Open	High	Low	Close	Volume
1-Jan-10	57.97	57.97	57.97	57.97	0
4-Jan-10	59.3	59.45	59.08	59.15	3956568
5-Jan-10	59.65	59.9	59.31	59.57	4109563
6-Jan-10	59.52	59.92	59.34	59.88	6230175
7-Jan-10	59.92	60	59.69	59.86	4432171
8-Jan-10	59.79	60.06	59.67	60	3786118
11-Jan-10	61.54	61.97	61.45	61.89	8229519
12-Jan-10	60.86	61.69	60.8	61.5	6127216
13-Jan-10	61.31	61.98	61	61.8	5688621
14-Jan-10	61.22	61.75	61.16	61.73	3330668
15-Jan-10	61.64	61.64	61.64	61.64	0
18-Jan-10	61.64	61.64	61.64	61.64	0
19-Jan-10	61.76	62.38	61.76	62.32	4914423
20-Jan-10	61.35	61.44	60.6	61.06	5410892
21-Jan-10	60.95	61.11	59.13	59.57	7184905
22-Jan-10	58.77	59.1	57.76	57.87	7297480
25-Jan-10	58.55	58.79	58.21	58.55	4041969
26-Jan-10	57.95	58.92	57.81	58.49	5685433
27-Jan-10	57.89	58.18	57.44	58.06	6026843

Descriptive Statistics	Value
Number of Data Points	259
Minimum Closing Price	27.02
Maximum Closing Price	62.32
Range (Max Closing price - Min Closing Price)	35.3
Average Closing Price	45.4
Standard Deviation of Closing Prices	9.391236824
Variance of the Closing Prices	88.19532908
Histogram	
number of bins	10
Bin size	3.53

Bin Minimum	Bin Maximum	Frequency
27	=H20+J14	

Figure A.44 The bin maximum limit is found by adding the bin size to the bin minimum limit.

chart should appear. Right-clicking on this chart should allow you to select the data source. Once the select data option is clicked, a window similar to the one shown in Figure A.47 should appear. This is a common window that appears for all chart types in Excel.

A

Close	Volume			Descriptive Statistics	Value
ANCE 2010					
Close	Volume				
57.97	0				
59.15	3956568			Descriptive Statistics	Value
59.57	4109563			Number of Data Points	259
59.88	6230175			Minimum Closing Price	27.02
59.86	4432171			Maximum Closing Price	62.32
60	3786118			Range (Max Closing price - Min Closing Price)	35.3
61.89	8229519			Average Closing Price	45.4
61.5	6127216			Standard Deviation of Closing Prices	9.391236824
61.8	5688621			Variance of the Closing Prices	88.19532908
61.73	3330668			Histogram	
61.64	0			number of bins	10
61.64	0			Bin size	3.53
62.32	4914423				
61.06	5410892				
59.57	7184905				
57.87	7297480				
58.55	4041969		**Bin Minimum**	**Bin Maximum**	**Frequency**
58.49	5685433		27	30.53	
58.06	6026843		=I20+0.01		
57.33	5003768				
56.12	7646893				

B

Date	Open	High	Low	Close	Volume			Descriptive Statistics	Value
BP NYSE STOCK PERFORMANCE 2010									
Date	Open	High	Low	Close	Volume				
1-Jan-10	57.97	57.97	57.97	57.97	0				
4-Jan-10	59.3	59.45	59.08	59.15	3956568			Descriptive Statistics	Value
5-Jan-10	59.65	59.9	59.31	59.57	4109563			Number of Data Points	259
6-Jan-10	59.52	59.92	59.34	59.88	6230175			Minimum Closing Price	27.02
7-Jan-10	59.92	60	59.69	59.86	4432171			Maximum Closing Price	62.32
8-Jan-10	59.79	60.06	59.67	60	3786118			Range (Max Closing price - Min Closing Price)	35.3
11-Jan-10	61.54	61.97	61.45	61.89	8229519			Average Closing Price	45.4
12-Jan-10	60.86	61.69	60.8	61.5	6127216			Standard Deviation of Closing Prices	9.391236824
13-Jan-10	61.31	61.98	61	61.8	5688621			Variance of the Closing Prices	88.19532908
14-Jan-10	61.22	61.75	61.16	61.73	3330668			Histogram	
15-Jan-10	61.64	61.64	61.64	61.64	0			number of bins	10
18-Jan-10	61.64	61.64	61.64	61.64	0			Bin size	3.53
19-Jan-10	61.76	62.38	61.76	62.32	4914423				
20-Jan-10	61.35	61.44	60.6	61.06	5410892				
21-Jan-10	60.95	61.11	59.13	59.57	7184905				
22-Jan-10	58.77	59.1	57.76	57.87	7297480				
25-Jan-10	58.55	58.79	58.21	58.55	4041969		Bin Minimum	Bin Maximum	Frequency
26-Jan-10	57.95	58.92	57.81	58.49	5685433		27	30.53	
27-Jan-10	57.89	58.18	57.44	58.06	6026843		30.54		
28-Jan-10	58.4	58.48	56.96	57.33	5003768				
29-Jan-10	57.08	57.38	55.87	56.12	7646893				

Figure A.45 The minimum bin limit of the next bin is obtained by simply adding a very small number (0.01 in this case) to the previous bin's maximum limit.

The following options are applicable for this graph:

1. Chart data range: Select the data stored in the Frequency column.
2. Name: Enter the name of the data series that is being plotted. Here the name of the data series is Frequency.
3. X values: Select the data stored in the Bin Maximum column as the X values data.
4. Y values: The data stored in the Frequency column automatically gets stored as the Y values data. This is because these data were selected as the Chart data range in option 1 above.

A

BP NYSE STOCK PERFORMANCE 2010

Date	Open	High	Low	Close	Volume
1-Jan-10	57.97	57.97	57.97	57.97	0
4-Jan-10	59.3	59.45	59.08	59.15	3956568
5-Jan-10	59.65	59.9	59.31	59.57	4109563
6-Jan-10	59.52	59.92	59.34	59.88	6230175
7-Jan-10	59.92	60	59.69	59.86	4432171
8-Jan-10	59.79	60.06	59.67	60	3786118
11-Jan-10	61.54	61.97	61.45	61.89	8229519
12-Jan-10	60.86	61.69	60.8	61.5	6127216
13-Jan-10	61.31	61.98	61	61.8	5688621
14-Jan-10	61.22	61.75	61.16	61.73	3330668
15-Jan-10	61.64	61.64	61.64	61.64	0
18-Jan-10	61.64	61.64	61.64	61.64	0
19-Jan-10	61.76	62.38	61.76	62.32	4914423
20-Jan-10	61.35	61.44	60.6	61.06	5410892
21-Jan-10	60.95	61.11	59.13	59.57	7184905
22-Jan-10	58.77	59.1	57.76	57.87	7297480
25-Jan-10	58.55	58.79	58.21	58.55	4041969
26-Jan-10	57.95	58.92	57.81	58.49	5685433
27-Jan-10	57.89	58.18	57.44	58.06	6026843
28-Jan-10	58.4	58.48	56.96	57.33	5003768
29-Jan-10	57.08	57.38	55.87	56.12	7646893
1-Feb-10	56.47	57.26	56.47	57.23	6036100
2-Feb-10	54.4	55.71	54.14	55.46	18575391
3-Feb-10	55.35	55.53	54.85	55.17	8413765
4-Feb-10	54.76	54.78	53.44	53.48	8366477
5-Feb-10	53.11	53.31	52.1	53.18	11067306
8-Feb-10	52.64	53.24	52.13	52.43	7811639
9-Feb-10	52.92	54.1	52.73	53.61	8017454

Descriptive Statistics	Value
Number of Data Points	259
Minimum Closing Price	27.02
Maximum Closing Price	62.32
Range (Max Closing price - Min Closing Price)	35.3
Average Closing Price	45.4
Standard Deviation of Closing Prices	9.391236824
Variance of the Closing Prices	88.19532908
Histogram	
number of bins	10
Bin size	3.53

Bin Minimum	Bin Maximum	Frequency
	27	30.53 =frequency(E3:E261,I20:I29)
30.54	34.07	
34.08	37.61	
37.62	41.15	
41.16	44.69	
44.7	48.23	
48.24	51.77	
51.78	55.31	
55.32	58.85	
58.86	62.39	

B

BP NYSE STOCK PERFORMANCE 2010

Date	Open	High	Low	Close	Volume
1-Jan-10	57.97	57.97	57.97	57.97	0
4-Jan-10	59.3	59.45	59.08	59.15	3956568
5-Jan-10	59.65	59.9	59.31	59.57	4109563
6-Jan-10	59.52	59.92	59.34	59.88	6230175
7-Jan-10	59.92	60	59.69	59.86	4432171
8-Jan-10	59.79	60.06	59.67	60	3786118
11-Jan-10	61.54	61.97	61.45	61.89	8229519
12-Jan-10	60.86	61.69	60.8	61.5	6127216
13-Jan-10	61.31	61.98	61	61.8	5688621
14-Jan-10	61.22	61.75	61.16	61.73	3330668
15-Jan-10	61.64	61.64	61.64	61.64	0
18-Jan-10	61.64	61.64	61.64	61.64	0
19-Jan-10	61.76	62.38	61.76	62.32	4914423
20-Jan-10	61.35	61.44	60.6	61.06	5410892
21-Jan-10	60.95	61.11	59.13	59.57	7184905
22-Jan-10	58.77	59.1	57.76	57.87	7297480
25-Jan-10	58.55	58.79	58.21	58.55	4041969
26-Jan-10	57.95	58.92	57.81	58.49	5685433
27-Jan-10	57.89	58.18	57.44	58.06	6026843
28-Jan-10	58.4	58.48	56.96	57.33	5003768
29-Jan-10	57.08	57.38	55.87	56.12	7646893
1-Feb-10	56.47	57.26	56.47	57.23	6036100
2-Feb-10	54.4	55.71	54.14	55.46	18575391
3-Feb-10	55.35	55.53	54.85	55.17	8413765
4-Feb-10	54.76	54.78	53.44	53.48	8366477
5-Feb-10	53.11	53.31	52.1	53.18	11067306
8-Feb-10	52.64	53.24	52.13	52.43	7811639
9-Feb-10	52.92	54.1	52.73	53.61	8017454

Descriptive Statistics	Value
Number of Data Points	259
Minimum Closing Price	27.02
Maximum Closing Price	62.32
Range (Max Closing price - Min Closing Price)	35.3
Average Closing Price	45.4
Standard Deviation of Closing Prices	9.391236824
Variance of the Closing Prices	88.19532908
Histogram	
number of bins	10
Bin size	3.53

Bin Minimum	Bin Maximum	Frequency
	27	30.53 12
30.54	34.07	
34.08	37.61	
37.62	41.15	
41.16	44.69	
44.7	48.23	
48.24	51.77	
51.78	55.31	
55.32	58.85	
58.86	62.39	

Figure A.46 Using the Excel function frequency() to determine the number of closing stock prices that belong to the first bin.

On clicking OK, the column chart displayed in Figure A.48 will appear. This chart needs to be formatted to prepare the final histogram. The first step in formatting the chart is to click on any one of the columns and right-click. On selecting the Format Data Series option, the Format Data Series window shown in Figure A.49 is displayed. Modify the Gap Width to 0% and use other options, including Excel's Quick Chart options, to create a histogram with the desired appearance.

A

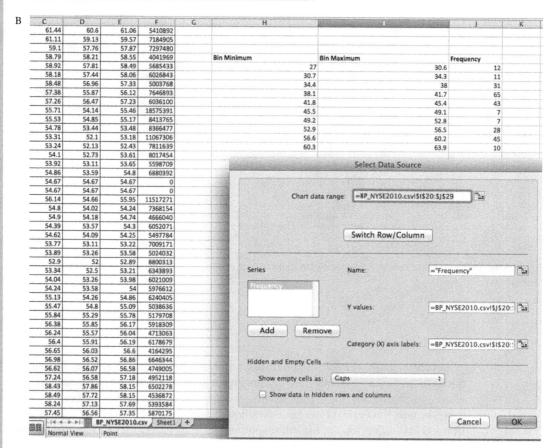

B

Figure A.47 The Select Data Source window appears when the Select Data option is clicked for an Excel chart.

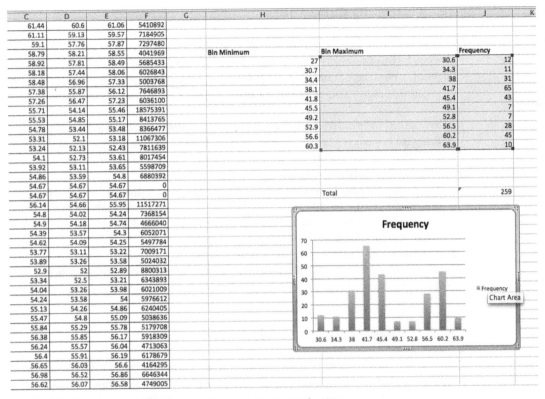

Figure A.48 Column chart to prepare the histogram for BP stock prices.

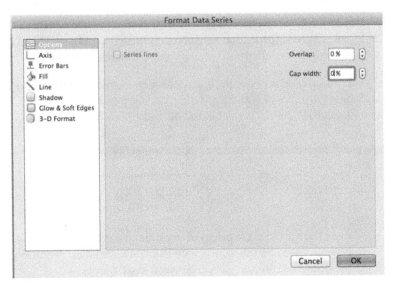

Figure A.49 The Format Data Series window allows users customize the column chart. Change the Gap width to 0%.

3.6   **Illustrating Correlations**

Correlation, as the name suggests, captures the relation between the behaviors of separate datasets. It gives a sense of how datasets behave together. Correlation between datasets can be positive, negative, or nonexistent. Let's assume that we have two datasets that represent the price of stock A and stock B. Any one of the following three cases is possible:

1. Two stocks are positively correlated: The price of stock A goes up at the same time the price of stock B goes up. Alternatively, this case can also be phrased as the price of stock B goes up at the same time the price of stock A goes up, or the stock prices of A and B are positively correlated.
2. Two stocks are negatively correlated: The price of stock A drops when the price of stock B goes up. Alternatively, this case can also be phrased as the price of stock B drops when the price of stock A goes up, or the stock prices of A and B are negatively correlated.
3. Two stocks are uncorrelated: There is no consistent pattern in the variation of the two stock prices. Alternatively, this case can also be phrased as the stock prices of A and B are uncorrelated.

3.6.1   **Pearson Correlation Coefficient**

The Pearson correlation coefficient is one of the most commonly used correlation coefficients. It indicates the degree of linear relationship between two datasets (stock prices).

The Pearson correlation coefficient is a unit-less quantity with values that vary between -1 and $+1$.

1. Perfect positive correlation between the datasets is indicated by $+1$.
2. Perfect negative correlation between the datasets is indicated by -1.
3. If the datasets are uncorrelated, the value of this coefficient is 0.

For stocks A and B, assuming that we have stock prices for n days (i.e., the count $= n$), and the stock prices of A and B are indicated by a_i and b_i where $i = 1 \ldots n$, the Pearson correlation coefficient R is calculated using the following equation:

$$R = \frac{\sum_{i=1}^{n}(a_i - \bar{a})(b_i - \bar{b})}{\sqrt{\sum_{i=1}^{n}(a_i - \bar{a})^2 \sum_{i=1}^{n}(b_i - \bar{b})^2}} \tag{4}$$

Here *a* and *b* are the average prices of stock A and stock B, respectively.

In general, based on the value of the Pearson correlation coefficient, we can determine the strength of the correlation between two datasets.

1. Strong Correlation: 0.75 to 1 and -0.75 to -1

2. Weak Correlation: 0.35 to 0.74 and −0.35 to −0.74
3. Very Weak or No Correlation: −0.34 to 0.34

Scatter plots can be used to visualize the correlation between datasets. For example, the scatter plots in Figure A.50 depict cases of perfect positive and negative correlation, respectively.

Figure A.50 (A) A scatter plot depicting perfect positive correlation (Pearson correlation coefficient = +1). (B) A scatter plot depicting perfect negative correlation (Pearson correlation coefficient = −1).

3.6.2 **Pearson Correlation Coefficient and Best-Fit Lines**

The general idea behind calculating the Pearson product-moment correlation of two datasets involves drawing a straight line that best fits the datasets—that is, a straight line that passes through the data points such that it is closer to all the points than any other straight line. Figure A.51 illustrates how a best-fit line is chosen. As can be seen in Figure A.51, several lines can be used to fit to the data, but the best-fit line is the one that minimizes the sum of the squares of the distances of the individual data points from the line.

Once the best-fit line is determined, the Pearson correlation coefficient r indicates how far these data points are from the best-fit line. The closer the data points from the best-fit line, the higher the correlation between them. The more spread out the data points (i.e., the farther away the data points from the best-fit line), the lower the correlation between them.

In Figure A.50, you can see that the best-fit lines pass through all the data points. This is because the data points are perfectly correlated (positively or negatively). The scatter plots also indicate the equation of the best-fit lines. Additionally, there is a quantity R^2 that is indicated underneath the best-fit line equation.

The R^2 value is calculated by squaring the Pearson correlation coefficient. R^2 quantifies the strength of the correlation. Since R^2 is the square of the Pearson correlation coefficient, it is always positive and hence does not provide an idea about whether the correlation is positive or negative. This can be noticed in both of the scatter plots of Figure A.50, where the R^2 value is positive and equal to 1.

Other examples of strong and weak correlation are provided in Figure A.52.

3.6.3 **Correlation in Excel**

Excel can be used to calculate the correlation values by implementing Equation (4). You can also use the correl(Array A, Array B) function in Excel, where Array A and Array B can be any two columns in Excel. Make sure that Array A and Array B have the same number of elements.

3.7 **Identifying Pearson Correlation Coefficient Limitations**

While the Pearson correlation coefficient can provide an insight into the correlation between datasets, it does have some limitations. Some of these limitations are discussed here.

3.7.1 **Linear Relationship**

The Pearson correlation coefficient uses the best-fit line—that is, it works with datasets that exhibit a linear relationship. If the datasets do not exhibit a linear relationship (an example of a nonlinear relationship between datasets is provided in Fig. A.53), the Pearson correlation coefficient should not be used.

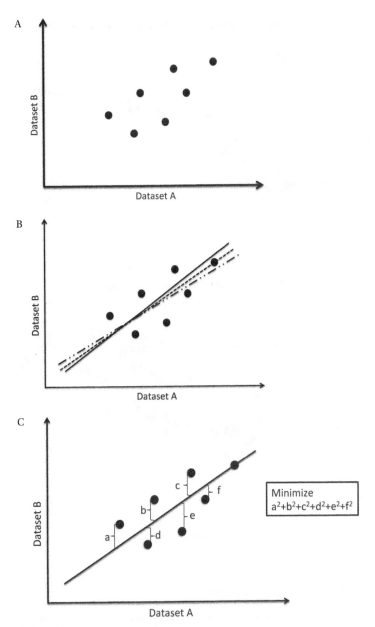

Figure A.51 Best-fit line generation process.

3.7.2 Slope Does Not Indicate Strength of the Correlation

The slope of the best-fit line does not indicate the strength of the correlation. This is demonstrated in Figure A.54, where the slope of the two scatter plots is different even though the datasets are perfectly correlated.

A

B

Figure A.52 The scatter plots illustrate different strengths of positive and negative correlation between the two datasets.

3.7.3   Correlation and Causation

An important point to note is that while correlation captures the collective behavior of datasets, it does not necessarily imply causation. In other words, high correlation between two variables does not mean that the behavior of one of them causes the other to behave in the same manner.

3.8   **Smoothing out Fluctuations of Data**

In many cases, datasets resulting from engineering systems consist of a series of data points whose values are recorded at successive intervals of time over a large period of

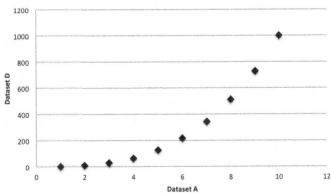

Figure A.53 Pearson correlation coefficient should not be used when the scatter plot indicates a nonlinear relationship between the datasets.

A

B

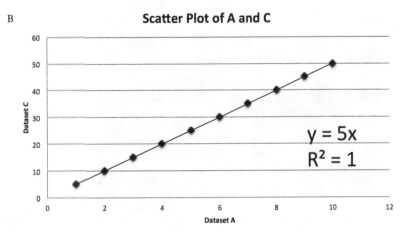

Figure A.54 The slope of the best-fit line does not provide information about the strength of the Pearson correlation coefficient. In this case, both datasets are perfectly correlated; however, the slopes of the best-fit lines are different.

time. Such datasets are referred to as a "time series." Examples of time series include environmental engineering data such as CO_2 emissions over time or temperatures over time, financial engineering data such as stock prices over time, joint angle data using motion capture systems, and sensor data such as accelerometer data used in rocket flights (Fig. A.55).

These time series have fluctuations in them that can be random or deterministic in nature. At times, there are patterns embedded in these fluctuations that can help us understand the behavior of the engineering process or system. Smoothing these fluctuations accentuates these patterns. Different types of smoothing functions have been developed to allow for smoothing the fluctuations out of data. Moving average is a useful concept that, when used effectively, can help smooth fluctuations that are inherent to data generated by several engineering processes and systems.

3.8.1 Simple Moving Average

A simple moving average (SMA), as the name suggests, computes the mean price of a stock over a specific number of periods called "taps." As new data points are measured, old data points are dropped. In this manner, the updated mean is calculated for a specific number of taps and the average moves forward in time.

Figure A.55 Time series are generated in several engineering systems. Here a motion capture system generates time series of human body joint angles that can be used to capture and analyze motion and render animations. http://upload.wikimedia.org/wikipedia/commons/7/73/ MotionCapture.jpg

3.8.2 Calculating SMA in Excel

Excel can be used to calculate a SMA easily. Consider the temperature of a room that is measured once a day for 10 consecutive days (Fig. A.56). To calculate a SMA (for example, a 3-tap SMA), we have to wait until the 3 temperature data points are measured and recorded. Once the 3 data points are recorded, calculating the SMA is simply a matter of using the average() function in Excel (Fig. A.56a).

As new data points are recorded, the SMA moves forward (Fig. A.56b). Here, the temperature at day 1 is discarded and the new temperature reading at day 4 is incorporated in the SMA. In this manner the SMA is calculated for all data points (Fig. A.56c).

The true benefit of a SMA is realized when it is used on large datasets that exhibit fluctuations. For example, the closing stock prices of Nanosphere, an American company that develops diagnostic solutions based on nanotechnology, is depicted in

A

	A	B
1	Temperature	Simple Moving Average
2	37	
3	32	
4	33	=AVERAGE(A2:A4)
5		
6		
7		
8		
9		
10		
11		

B

	A	B
1	Temperature	Simple Moving Average
2	37	
3	32	
4	33	34
5	38	=average(A3:A5)
6		
7		
8		
9		
10		
11		

C

	A	B
1	Temperature	Simple Moving Average
2	37	
3	32	
4	33	34
5	38	34.33333333
6	31	34
7	34	34.33333333
8	36	33.66666667
9	30	33.33333333
10	39	35
11	35	=AVERAGE(A9:A11)

Figure A.56 An example of a 3-tap SMA calculation for 10-day temperature time series. (A) To calculate the first SMA value, one has to wait until 3 measurements are available. (B) As new points are available, old ones are discarded and the average is calculated again. (C) This process is repeated until all points are accounted for.

Figure A.57. It is clear that despite all the fluctuations, the overall trend in the stock prices between October 2011 and October 2012 was that of mild growth, slow decline, and rapid growth, respectively (the three arrows in Fig. A.57). Notice how the 24-day SMA starts 24 days after the first closing price is recorded.

3.8.3 Lag in Moving Averages

Since SMA is based on past data points, we have to wait until these past data points are recorded. This results in a lag in the moving average. The more taps in the moving average, the more lag it will have in relation to the current behavior of the engineering system or process. The shorter the moving average, the more sensitive it is to the fluctuations of the data. The more taps in a moving average, the less sensitive it will be to the fluctuations. However, moving averages with more taps are better highlighters of trends embedded in the data.

An example of the lag is clearly visible in Figure A.58. Here we show a 10-day SMA and a 24-day SMA. Clearly, the 10-day SMA is more sensitive (and hence there is less lag) than the 24-day SMA. However, the 24-day SMA clearly captures the overall trends, whereas the 10-day SMA fluctuates more.

3.9 **Using Line Graphs and Scatter Plots**

Visualization of data is a critical part of data analytics. Visualizations become increasingly important as the volume of data stored in Excel grows. Excel can be used to visualize data using the plethora of chart options that it provides.

Figure A.57 A 24-day SMA of Nanosphere closing prices when superimposed on the actual closing prices clearly accentuates the stock price trends.

Figure A.58 10-day and 24-day SMAs of Nanosphere closing prices, when superimposed on the actual closing prices, clearly reveal that the 10-day SMA is more sensitive (and hence there is less lag) to the fluctuations, whereas the 24-day SMA is less sensitive to the fluctuations.

3.9.1 Line Graphs

Line graphs are an important class of visualizations. These graphs present information in the form of a series of data points connected by straight-line segments. They represent how a series of data points change with respect to time or any other variable.

Referring to data shown in Figure A.11, let's plot the columns X2 and Y as a line graph. To create the line plot, select the column Y (including the first cell that contains the name of the column). Next, select the line graph option provided in the Charts menu of Excel. This will generate the graph shown in Figure A.59.

To plot X2 and Y on the same line graph, select X2 and Y (including the first cells that contain the column names). Subsequent to this, select the line graph option provided in the Charts menu of Excel. This will generate the graph shown in Figure A.60.

3.9.2 Scatter Plots

Scatter plots help visualize correlation between datasets. Correlation, as the name suggests, captures the relation between the behaviors of separate datasets. It gives a sense of how datasets behave together. Correlation was discussed in detail in Section 3.6 of this appendix.

Scatter plots can be two-dimensional or three-dimensional. We will focus on two-dimensional scatter plots. Consider two datasets described in the columns A and B of Figure A.61. A scatter plot for these two columns can be plotted by selecting the two columns and using the Scatter Plot option in the Charts menu of Excel.

Similarly, as shown in Figure A.62, a scatter plot with datasets A and C is displayed.

Figure A.59 A line graph of column Y. Notice how Excel provides a title and legend for this. The axes are not labeled yet. For a complete graph, the axes need to be labeled.

Figure A.60 A line graph of columns X2 and Y on the same figure. Notice how Excel does not provides a title in this case. The legends are provided. The axes are not labeled yet.

While a few examples of visualizations are discussed here, readers are encouraged to explore more chart options provided in the Charts menu in Excel.

A.4   **Critical Analysis and Data Analytics: Practice Projects**

Data analytics requires critically analyzing a dataset by asking fundamental questions about the problem at hand. The four stages of data analytics presented in Figure A.63 are described in Section 2.

Figure A.61 A scatter plot for columns A and B. In this case, we have shown the title and axes labels. As can be seen here, the scatter plot reveals that as the value of A increases, the values stored in B also increase. This means that the data stored in columns A and B are positively correlated.

Figure A.62 A scatter plot for columns A and C. In this case, we have shown the title and axes labels. As can be seen here, the scatter plot reveals that as the value of A increases, the values stored in C decreases. This means that the data stored in columns A and B are negatively correlated.

1. Procuring Data: Data can be procured from various sources such as websites, software database, or sensors. Data can also be manually entered into Excel. Not all the rows and columns in a dataset are important, so carefully select which columns will be analyzed. You can delete a column or hide it using Excel's Delete or Hide commands. See Section 5 for more details.

2. Analysis: Critically analyzing data requires answering questions about the trends, distributions, fluctuations, and correlations. Figure A.64 illustrates several aspects of performing analyses of datasets and visualizing the results. Aspects such as trends are important to understand the past behavior of the data points or systems that generate the datasets. Analyzing and visualizing the distribution allow us to gain an understanding of central tendency and spread of the data. If the data points

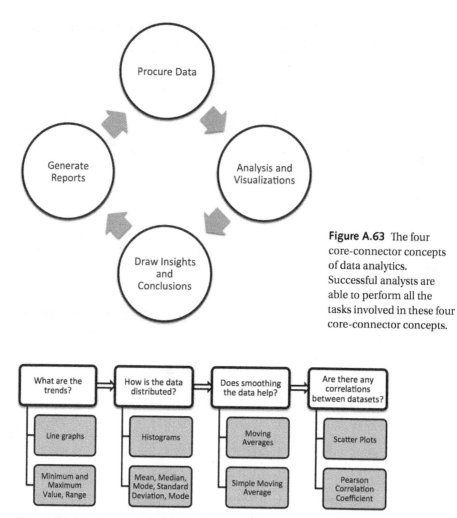

Figure A.63 The four core-connector concepts of data analytics. Successful analysts are able to perform all the tasks involved in these four core-connector concepts.

Figure A.64 Key aspects of critically analyzing datasets.

feature fluctuations, we can use SMAs to smooth these fluctuations and study any trends that emerge. Finally, when multiple datasets are available, it is useful to understand and measure any correlations between them.

3. Draw Insights: Once the data have been statistically analyzed, we try to glean insights and draw conclusions about the datasets or the engineering systems and processes that generated them. Insights can be gleaned by inspecting the visualizations and the actual values. Conclusions can be drawn by synthesizing the insights with any other relevant information.

4. Generate Reports: Finally, we need to generate reports to describe the key insights and conclusions. Reports can be text files or webpages or any other format required by the reader. In any report that describes data analytics, we must strive to find a

healthy balance between descriptive text and visualizations. The key is to provide just enough background information to the reader so that he or she can interpret the visualizations correctly.

In the subsequent sections, we will critically analyze datasets from financial and environmental engineering.

A.5 Practice Project 1: Analyzing Environmental Engineering Data

Environmental engineers are professionals who analyze, design, construct, manage, or operate facilities or systems to protect or enhance the environment of people and other living things, or advance knowledge of the field. [1] A large focus of environmental engineering is to find innovative solutions to managing pollution.

The Environmental Protection Agency (EPA) is an agency of the U.S. federal government that was created to protect human health and the environment by writing and enforcing regulations based on laws passed by Congress. The EPA's website hosts a large amount of data pertaining to emissions of greenhouse gases such as carbon dioxide. In the subsequent sections we will analyze environmental data pertaining to CO_2 emissions and temperatures.

5.1 A Critical Analysis of Global Trends in CO_2 Emissions

Studying global CO_2 emissions is critical to our understanding of the trends in global pollution. Critically analyzing the CO_2 emission data can provide insights about several important questions. These questions can solicit information about the following:

1. Country with the most and least average CO_2 emissions
2. The percentage of emissions resulting from BRIC (Brazil, Russia, India, and China) nations in a given year
3. Comparison of the U.S. and China CO_2 emissions since 1990
4. Major emitters in the continent of Africa

5.2 Procuring Emissions Data

There are two organizations that provide CO_2 emission data:

1. Carbon Dioxide Information Analysis Center (CDIAC) (http://cdiac.ornl.gov/)
2. United Nations Framework Convention on Climate Change (http://unfccc .int/2860.php)[2]

Here we will work with data provided by the Millennium Development Goal Indicators website of the United Nations.

---

[1] http://www.drexel.edu/cae/about/objectives/

[2] Download: http://mdgs.un.org/unsd/mdg/Data.aspx Goal7. Target 7A.

Figure A.65 provides a snapshot of the webpage that hosts the dataset we will use for our analysis. This dataset can be downloaded as an Excel file by clicking on the Excel logo provided on the webpage. Once downloaded and opened in Excel, the dataset has

Figure A.65 The UN MDG website provides the required data for this analysis. Data can be downloaded in Excel format by clicking on the Excel button on the top right of the webpage.

	B	C	D	E	F	G
1	7.2 Carbon dioxide emissions, total, per capita and per $1 GDP (PPP)					
2	Country	1990	1990 Nature	1990 Footnote	1991	1991 Nature
3	Carbon dioxide emissions (CO2), thousand metric tons of CO2 (CDIAC)					
4	Afghanistan	2677	G		2494	G
5	Albania	7488	G		3971	G
6	Algeria	78896	G		81459	G
7	Andorra					
8	Angola	4430	G		4367	G
9	Anguilla					
10	Antigua and Ba	301	G		293	G
11	Argentina	112614	G		117021	G
12	Armenia					
13	Aruba	1841	G		1929	G
14	Australia	287331	G		281530	G
15	Austria	60726	G		65603	G
16	Azerbaijan					
17	Bahamas	1951	G		1782	G
18	Bahrain	11885	G		11503	G
19	Bangladesh	15533	G		15940	G
20	Barbados	1074	G		1206	G
21	Belarus					
22	Belgium	108470	G		117579	G
23	Belize	312	G		359	G

Figure A.66 The UN MDG CO_2 emission data have been downloaded and opened in Excel. Several columns are either empty or have non-numeric characters that will not be used for the analysis.

several columns that are empty or have non-numeric characters (Fig. A.66). These columns will not be used in our analysis and hence should be either deleted or hidden using Excel's Delete or Hide commands. Selecting the two columns and right-clicking on them displays a menu with the Delete and Hide commands, as shown in Figure A.67. Figure A.68 depicts the dataset once these columns have been deleted.

Figure A.67 The columns that will not be used for the analysis can be deleted or hidden using the Delete and Hide commands, respectively.

	B	C	D	E	F	G	H
1	Country	1990	1991	1992	1993	1994	
2	Carbon dioxide emissions (CO2), thousand metric tons of CO2 (CDIAC)						
3	Afghanistan	2677	2494	1426	1375	1320	
4	Albania	7488	3971	2387	2343	1929	
5	Algeria	78896	81459	82262	82474	86351	⁹
6	Andorra						
7	Angola	4430	4367	4419	5801	3891	⁻
8	Anguilla						
9	Antigua and Barbuda	301	293	293	308	312	
10	Argentina	112614	117021	121447	118609	123351	12
11	Armenia			4052	2897	2967	
12	Aruba	1841	1929	1723	1771	1764	
13	Australia	287331	281530	294456	302117	303958	30
14	Austria	60726	65603	60542	60282	61103	6
15	Azerbaijan			57682	49365	42673	⁻
16	Bahamas	1951	1782	1793	1716	1720	
17	Bahrain	11885	11503	10590	15141	15255	1
18	Bangladesh	15533	15940	17748	17407	18969	2
19	Barbados	1074	1206	979	1115	748	
20	Belarus			87487	74451	65914	6
21	Belgium	108470	117579	114374	108925	113307	11
22	Belize	312	359	356	378	374	
23	Benin	715	829	906	1133	1265	

Figure A.68 Once all the columns that will not be used for analysis have been deleted, the dataset is ready for statistical analysis.

5.3   Analyzing and Visualizing Emissions Data

Since the dataset has many columns and rows, it is worth noting that we can hide the columns to facilitate the analysis. As shown in Figure A.69, a simple analysis shows that the United States has emitted the most CO_2 on average, totaling 5443383.1 thousand metric tons per year between 1990 and 2009. Niue shows on average 4 thousand metric tons per year between 1990 and 2009.

A

	B	W	X	Y	Z	AA	AB	AC
1	Country		Average CO2 data					
2	Carbon dioxide emissions (CO2), thousand metric tons of CO2 (CDIAC)							
3	Afghanistan		1605.7					
4	Albania		3368.5					
5	Algeria		94749.5					
6	Andorra		512.2666667					
7	Angola		12365.45					
8	Anguilla		41.53846154					
9	Antigua and Barbuda		360.1					
10	Argentina		140309.1					
11	Armenia		3707.777778					
12	Aruba		2027.65					
13	Australia		336138.7					
14	Austria		65638.4					
15	Azerbaijan		37579.83333					
16	Bahamas		1976.25			Analysis		Answer
17	Bahrain		17081.55			Max Average Emission		=MAX(X3:X220)
18	Bangladesh		30311.7					
19	Barbados		1174					
20	Belarus		60975.72222					
21	Belgium		111852.55					
22	Belize		419.4					
23	Benin		2089.75					

Figure A.69 (a) Finding the maximum average emission value

B

1	Country		Average CO2 data					
2	Carbon dioxide emissions (CO2), thousand metric tons of CO2 (CDIAC)							
3	Afghanistan		1605.7					
4	Albania		3368.5					
5	Algeria		94749.5					
6	Andorra		512.2666667					
7	Angola		12365.45					
8	Anguilla		41.53846154					
9	Antigua and Barbuda		360.1					
10	Argentina		140309.1					
11	Armenia		3707.777778					
12	Aruba		2027.65					
13	Australia		336138.7					
14	Austria		65638.4					
15	Azerbaijan		37579.83333					
16	Bahamas		1976.25			Analysis		Answer
17	Bahrain		17081.55			Max Average Emission		5443383.1
18	Bangladesh		30311.7			Min Average Emission		=min(X3:X220)
19	Barbados		1174					
20	Belarus		60975.72222					
21	Belgium		111852.55					
22	Belize		419.4					
23	Benin		2089.75					

Figure A.69 (b) Finding the minimum average emission value

C

	B	W	X	Y	Z	AA	AB	AC	AD
1	Country		Average CO2 data						
2	Carbon dioxide emissions (CO2), thousand metric tons of CO2 (CDIAC)								
3	Afghanistan		1605.7						
4	Albania		3368.5						
5	Algeria		94749.5						
6	Andorra		512.2666667						
7	Angola		12365.45						
8	Anguilla		41.53846154						
9	Antigua and Barbuda		360.1						
10	Argentina		140309.1						
11	Armenia		3707.777778						
12	Aruba		2027.65						
13	Australia		336138.7						
14	Austria		65638.4						
15	Azerbaijan		37579.83333						
16	Bahamas		1976.25				Analysis	Value	Country
17	Bahrain		17081.55				Max Average Emission	5443383.1	United States of America
18	Bangladesh		30311.7				Min Average Emission	4	Niue
19	Barbados		1174						
20	Belarus		60975.72222						
21	Belgium		111852.55						
22	Belize		419.4						
23	Benin		2089.75						

Figure A.69 (c) A simple inspection reveals that the United States and Niue have the maximum and minimum average value.

5.4   Insights About Global Emissions

Figure A.70 indicates the growth in the global CO_2 emissions since 1990. As can be seen clearly, while there was a sharp increase in emissions during 1991–1992, the growth slowed between 1993 and 2002. However, starting in 2003, overall emissions have clearly risen dramatically. There are several reasons for this growth, the BRIC nations being one of them.

5.4.1   BRIC Nations

The BRIC nations (Brazil, Russia, India, and China) are considered to be at a similar stage of economic growth and development. The strong economic growth and development has resulted in a substantial increase in emissions produced by these countries. The pie chart in Figure A.71 represents the percentage of emissions resulting from the BRIC nations in 2007. It is clear that these nations emitted approximately 35% of the global emissions. China appears to be a major driver of these emissions, with it accounting for 17% of the global emissions (more than the combined emissions of Brazil, Russia, and India).

5.4.2   United States and China

The United States was the leading CO_2 emitter for many years. While the United States emitted the highest amount of CO_2 on average between 1990 and 2009, China has recently overtaken the United States in terms of actual CO_2. China is one of the fastest-growing economies, and the growth in emissions has been largely attributed to the rapid increase of manufacturing industries in China. This is clearly seen in Figure A.72.

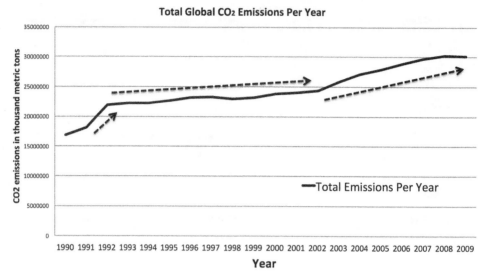

Figure A.70 Global CO_2 emissions have risen steadily since 1990.

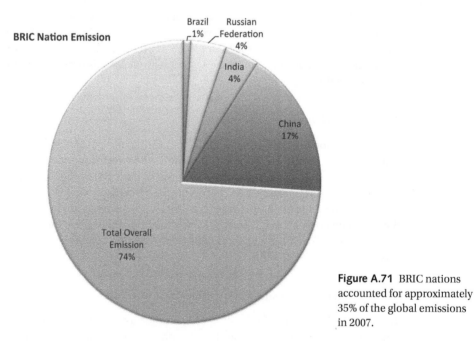

Figure A.71 BRIC nations accounted for approximately 35% of the global emissions in 2007.

5.4.3 African Nations

The African nations have shown a steady increase in CO_2 emissions since 1990. Countries such as South Africa, Egypt, Nigeria, and Libya are major emitters in Africa when

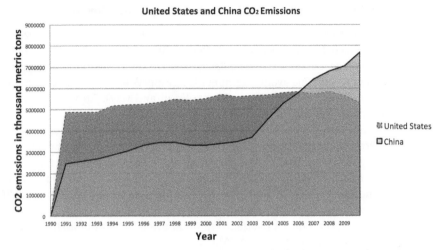

Figure A.72 China has surpassed the United States as the leading emitter of CO_2.

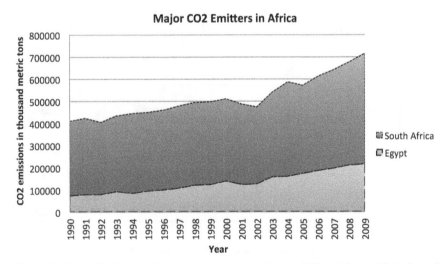

Figure A.73 South Africa and Egypt are two major emitters of CO_2 in Africa, with both totaling at approximately 700,000 thousand metric tons of CO_2 in 2009.

compared to other African nations. The two main emitters are South Africa and Egypt. As can be seen in Figure A.73, these two countries emitted approximately 700,000 thousand metric tons of CO_2 in 2009.

5.5   A Critical Analysis of New Jersey Temperature Trends

Let's look at New Jersey temperatures over the past few years. [3] Analyzing the temperature data of New Jersey reveals interesting insights about the hottest and coolest month on average, the hottest and coolest year on average, the month and year with the least and most variable average temperature respectively, and seasonal temperatures (e.g., spring season in the months of March, April, and May). Furthermore, it reveals a distinguishable trend in the average annual temperatures.

New Jersey witnessed its highest average annual temperatures in 1998, when the average temperature was 55.63°F. The coolest average annual temperatures were observed in 1904 (48.65°F). Winters are the most variable season in New Jersey, displaying a standard deviation of 2.68°F, whereas summers have remained fairly constant, with a standard deviation of 1.41°F. However, there is a positive correlation (0.304, Fig. A.74) between summers and winters in New Jersey.

Spring seasons are pleasant—never above 60°F—but more than half of the springs have been above 50°F (Fig. A.75).

In general, winters have been getting warmer, as is clearly observed by the 5-day SMA and EMA shown in Figure A.76. The coolest month on average is January, whereas the hottest month on average is July.

Figure A.74 A scatter plot of the correlation between New Jersey average winter and summer temperatures.

---

[3] The NJ temperature data can be downloaded from here: http://climate.rutgers.edu/stateclim_v1/data/njhisttemp.html

Store this data in a text file. Import the text file into Excel.

Figure A.75 A histogram illustrating the distribution of New Jersey spring temperatures

Figure A.76 A line graph of New Jersey seasonal average temperatures trends.

A.6 Practice Project 2: Analyzing Financial Engineering Data

Financial engineering is a multidisciplinary field involving financial theory, the methods of engineering, the tools of mathematics, and the practice of programming. [4] The complex interplay between technology and engineering innovation and operations, and financial growth requires students to have basic skills to analyze financial engineering datasets.

[4] http://ieor.columbia.edu/ms- financial- engineering

6.1   Apple Inc.: Line Graphs, Engineering Innovations, Financial Growth

Apple Inc. is a multinational technology company that was formed in 1976–1977. It has introduced a host of innovative products in the market, most notably the iPhone and the iPad. Apple witnessed its most profitable years starting in 2004–2005. This bullish trend is evident in the NASDAQ closing prices of Apple stock shown in Figure A.77. During the Worldwide Developer Conference (WWDC) held in January 2005, Steve Jobs announced that Apple would use Intel processors for its Macintosh computers running Mac OS X. This move was an effort toward transitioning the Macintosh line of computers to Intel-powered processors. On December 13, 2005, Apple's stock closing price reached a new high, $74.98.

In 2007, Apple announced and launched the first iPhone. The iPhone disrupted the smartphone industry and accelerated Apple's revenue growth. Apple's stock closing price on the day of the iPhone launch was $122.04, or 235% higher than the closing price on the same date in 2005. By the end of 2007, the stock price reached $198.08, a 62% increase since the launch of the iPhone. In 2010, Apple launched the first iPad. The tablet registered strong sales of 300,000 on the first day of its release.[5] This resulted in a strong Apple stock performance. The closing price around the time of launch was approximately $235, and by the end of 2010, the stock price was $322, a 37% increase since the launch of the iPad. In October 2011, Steve Jobs passed away. Between November 2011 and April 2012, the average closing price of Apple stock has registered its fastest growth, from approximately $385 to $606, a 57% increase. This rapid growth is primarily attributed to strong iPhone and iPad sales.[6]

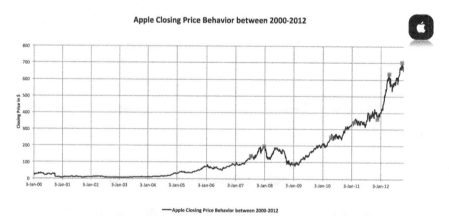

Figure A.77 Apple has witnessed strong growth starting 2004–2005. The NASDAQ closing prices of Apple stocks clearly indicate this growth.

[5] Apple Sells Over 300,000 iPads First Day, Official press release. Available here: http://www.apple.com/pr/library/2010/04/05Apple-Sells-Over-300-000-iPads-First-Day.html

[6] Apple Reports Third Quarter Results, Official press release. Available here: http://www.apple.com/pr/library/2012/07/24Apple-Reports-Third-Quarter-Results.html

6.2 IBM, Nanotechnology, Histograms, and Moving Averages

IBM is a multinational company that has built a reputation for quality, reliability, and the ability to operate profitably in good times and bad (Figs. A.78 and A.79).[7] The moving averages accentuate several long-term bullish trends in IBM stock prices, as indicated by the green dashed arrows in the line graph in Figure A.79. These bullish trends have spanned up to 7 years (1980–1987). On the other hand, the bearish trends are infrequent and relatively short-lived, as indicated by the red dashed arrows. Of special note is the fact that during all these periods, the average closing price has remained between $100 and $114.25. As seen from the histogram in Figure A.78, approximately 66% of the time since 1980, the stock has remained between $80.50 and $139.82, while the stock dipped below this range only 20% of the time. Such long-term, stable behavior allows investors to trade low volumes of IBM stock at high profit margins per share. Hence, IBM could be part of a long-term investment plan.

Nanosphere is an American company that develops diagnostic solutions based on nanotechnology. Its stock has traded between $1.27 and $3.59 since October 2011. The stock closed at an average of $2.19. As clearly indicated by the moving averages in Figure A.80, the stock prices have risen since 2011 and display bullish and bearish trends in quick succession. This sporadic behavior is also seen in the histogram in Figure A.81, which indicates that the stock has performed above average about 20% of the time and below average approximately 60% of the time. These low prices, coupled with such sporadic behavior, make this stock suitable for high-volume, short-term trades.

IBM Stock Closing Price Distribution 1980-2012

	60.762	80.534	100.297	120.06	139.823	159.586	179.349	199.112	218.875	238.638	258.401
Frequency	147	203	399	444	296	82	73	43	22	1	1

Figure A.78 IBM stock distribution.

[7] http://www.nyse.com/content/faqs/1042235995602.html?cat=Listed_Company___ General

Figure A.79 IBM stock performance trends.

Figure A.80 Nanosphere stock performance trends.

Figure A.81 Nanosphere stock distribution.

A.7 Exercise Questions

1. Netflix stock in 2011

   (a) Use Yahoo/Google Finance to download the daily prices of Netflix stock for 2011 as a .csv file. Import the Netflix stock prices into the same Microsoft Excel file.

   (b) Use an Excel function to ascertain the number of data points for these stocks.

   (c) Use Excel functions to calculate the minimum and maximum prices of these stocks. What is the range of the stock prices?

   (d) Use an Excel function to find the average value of these stocks. What is the standard deviation of the stock prices?

   (e) Plot the stock prices as line graphs. Each graph should have clearly labeled axes and a title.

   (f) Create a histogram of the stock prices. Label the histogram with the necessary information.

   (g) Identify 3 significant changes in the stock behavior of Netflix (2011). Highlight these changes on the line graphs. Provide reasons for these changes. Compile a one-page document to present your findings and the annotated line graphs.

2. Perform the required data analytics to identify any trends in Microsoft stock performance when Windows 95, 98, XP, and Vista were released.

3. Through the Kyoto Protocol, a number of industrialized nations (called the Annex-I nations) entered into an agreement to reduce the overall CO_2 emissions. Using the CO_2 emissions data used in Section 5, answer the following questions.

    (a) Did the European Union nations in the Annex-I group of the Kyoto Protocol meet their targets? Why? Provide two reasons that support your conclusions. Provide citations for your sources of information.

    (b) Did Japan and the U.S. meet their targets? Why? Provide two reasons that support your conclusions. Provide citations for your sources of information.

    More information on Kyoto Protocol:

    http://unfccc.int/parties_and_observers/parties/annex_i/items/ 2774.php and http:// unfccc.int/resource/docs/publications/08_unfccc_kp_ref_ manual.pdf (page 13)

4. Select a city or a state and critically analyze the historical temperature trends for the last 10 years. For example, Philadelphia temperature datasets are available at http://www.fi.edu/weather/data2/index.html.

5. Let's focus our attention to the transportation sector of the U.S.

    (a) Using data from http://www.epa.gov/otaq/fetrends.htm, highlight the vehicle fuel economy trends in the U.S. from 1975–2011.

    (b) Highlight the trends in CO_2 emissions in the U.S. due to vehicle emissions from 1975–2011.

    (c) Is there a correlation between fuel economy and CO_2 emissions? Explain your findings.

    (d) Given your findings about vehicle fuel economy and CO_2 emissions in the U.S., propose one improvement to reduce CO_2 emissions due to the transportation sector. In your answer, discuss the rationale and pros and cons of your suggested solution. Please cite your sources of information.

APPENDIX B

The Basics of PowerPoint

B.1 Introduction

It is assumed that the user is familiar with general Windows methods of menu use, selecting of files, text entry and editing, drag-and-drop, etc. The Copy, Paste, and Cut features of the Edit Menu of Windows work within PowerPoint.

The information presented here is very basic but should be enough to prepare a simple PowerPoint presentation. This is in keeping with the "20-80 Rule for Software"; namely, that if you learn to do 20 key procedures within a software package, you will be able to do 80% of anything you will ever want to do with the software. Here you learn about:

  ... choosing a first slide,
  ... adding a background to that slide or all slides of the presentation,
  ... adding and arranging pictures or clipart on a slide,
  ... adding and editing basic text on a slide,
  ... getting another blank slide to work on,
  ... arranging slides that have been made, and
  ... displaying slides as a show.

The basics are all here but many bells and whistles that are available are not discussed. This includes adding of notes, charts, tables, geometric shapes, movies, and sounds to slides. Once the basic procedure is learned, users are urged to experiment with the many Menu Bar items and Toolbar buttons (particularly those in the Draw toolbar) that are available to see what can be discovered.

For additional information about using PowerPoint, the following sources might be useful:

1. Cliffsnotes at www.cliffsnotes.com offers publications about several versions of PowerPoint.
2. Enter "PowerPoint tutorials" or "PowerPoint books" into a search engine such as Google. This will yield a huge number of internet sites as potential resources. Some are selling related items but many have free online information.

B.2   **The Basics of PowerPoint**

1. Open the program and choose the "**Blank Presentation**" option.
   *Note: For convenience, activate the following toolbars: **Standard, Formatting, Drawing, Picture, WordArt**. To do so, right click anywhere on the Toolbar area and click on these items if a check mark does not already appear beside them. To identify what an icon button represents, move the mouse to it and wait. Only a few of the buttons that appear are referenced in the following. Experiment to see what others will do.*

2. Select a slide type and display your first slide. [It is advisable to begin saving your work at this point. Use **File-Save As** and give the new file (which will have a .ppt extension) a name and a location for storing it. Then, save periodically as you create new slides.]

3. If you would like a background for your slides, choose from the Menu bar the sequence **Format-Background.** A dialog box will appear from which you can choose a solid color and various fill patterns. The background will always remain behind any text or pictures you might add. The background can be uniform for all slides or individualized.

4. If a text block is included in the slide type chosen, click on it and enter the text you want. Click outside of the block to remove the block's border. To reactivate for editing, click on the text.

   . . .   Edit as you would any word processor text.
   . . .   You can drag and drop the text block to reposition it.
   . . .   You can use the Free Rotate button on the WordArt toolbar to rotate the text block.
   . . .   You can change the shape of the block by dragging any of its 8 handles (the small squares).
   . . .   You can change the text characteristics (font, size, bold, center, etc.) using toolbar options. You must highlight the text to which the changes are to apply.
   . . .   Right click on a text block for a menu of manipulations pertaining to it.

5. If additional text is desired, click on the "**Text Box**" button, or from the Menu Bar choose **Insert-Text Box**. Drag-and-drop the mouse to form a text block, which can be used as above.

6. Text in WordArt format can be added. Click on the "**Insert WordArt**" button on the Toolbar, select the desired format, enter the text, resize it, and reshape it if you wish. Explore the Toolbar buttons for manipulations that can be applied to WordArt text as well as regular text.

7. To import a picture file, click on the "**Insert Picture From File**" button on the toolbar or choose from the Menu bar the sequence **Insert-Picture-From File**. The small squares are the "handles."

   . . .   To resize the picture, move the mouse to any of the 8 handles and drag to the desired size.
   . . .   To move the picture, place the mouse inside the picture; when a 4-headed arrow set appears, drag and drop the picture to the desired location.

... To resize or reposition a picture if the handles are not showing, first click on the picture to reveal the handles and then proceed as above.

... Buttons on the Picture Toolbar allow you to crop the picture (choose the **Crop** button and move any handle) and to change the contrast and brightness of it.

... Right click on a picture for a menu of manipulations pertaining to it.

8. Clip art can be added by clicking on the "Insert Clip Art" toolbar icon, or using the Menu bar sequence Insert-Picture-ClipArt. Procedures for resizing and repositioning are the same as for pictures.

9. When a new item (text, picture, clip art, etc.) is added, it will be on top of anything already there. To change the visibility of an item, select the item to be moved by clicking on any visible part of it. Then, right click on the selected item, click on Order, and choose the destination of the selected item.

10. To delete any item at any time, display its handles (click on it) and press **Delete** or **Backspace.**

11. To begin working on a new slide, click on the "**New Slide**" button of the Toolbar or use **Insert-New Slide** from the Menu bar. Select a slide format and proceed as with the first slide.

12. After you have created several slides, you can view them all in thumbnail form by clicking on the small button "**Slide Sorter View**," which is one of the buttons in the lower left corner just above any toolbars that might be there.

... You can change the arrangement of the slides by dragging and dropping any slide into any other position. A vertical line indicates the location of the slide as you drag it.

... To bring a slide back to the full screen for editing or additional work, double click on it.

... To delete a slide, select the slide and press **Delete**.

... To copy a slide, select it and use **Edit-Duplicate**.

13. Be sure you have saved your work.

14. To display your slide show:

... If there are slides in the slide sorter view you do not want to include in the slide show, click on the slide to select it. Then, use the Menu bar sequence **Slide Show-Hide Slide** to mark it as hidden.

... Use **Slide Show-Set Up Show** to choose options on how the show will be presented.

... Use **Slide Show-Slide Transition** to choose a visual and/or audio transition pattern from slide to slide, if desired. A drop-down menu offers numerous transitions that can be applied individually or universally.

... If your slide contains text, pictures, or clip art, you can assign an animation to each specifying the manner in which they appear on the screen. Select the item to be animated by clicking on it, then use **Slide Show-Preset Animation,** and then the available type of animation to be used.

...   If you assign multiple animations to a slide, you can change the order of the animations by using **Slide Show-Custom Animation**. Highlight a line in the "**Animation Order**" box and use arrows at the right to change the order of the animations.

...   To view the show, use **Slide Show-View Show**. Click the mouse to progress from slide to slide if automatic advancement is not in effect.

...   If you press **Ctrl** and the letter **P** at the same time, you will have a "pen" that can be used to draw on the current slide by clicking and moving the mouse. (Marks are not permanent; the slide itself is unaffected.)

Note: PowerPoint makes an excellent layout for printing of a single picture or a simple collage of pictures. Prepare one slide with the format and content that you want to print, and then print that slide using **File-Print**.

Index

Page numbers in bold indicate figures, italic indicate tables.